Highly Integrated Low-Power Radars

Sergio Saponara

Maria Greco

Egidio Ragonese

Giuseppe Palmisano

Bruno Neri

ARTECH
HOUSE

BOSTON | LONDON
artechhouse.com

Library of Congress Cataloging-in-Publication Data
A catalog record for this book is available from the U.S. Library of Congress.

British Library Cataloguing in Publication Data
A catalogue record for this book is available from the British Library.

Cover design by Igor Valdman

ISBN 13: 978-1-60807-665-9

685 Canton Street
Norwood, MA 02062

10 9 8 7 6 5 4 3 2 1

Contents

Preface

The main idea behind this book is that radar, properly designed to minimize its power consumption, size, and cost, has the potential to become in the near future a ubiquitous contactless sensor for large market applications.

The large amount of business that can be derived from ubiquitous radar sensing justifies research from industry and academia in this direction.

The origin of this book was the plenary talk entitled "Advances in Technologies and Architectures for Low-Power and Highly-Integrated Ubiquitous Radars" that I was invited to give at the IEEE Radar Conference (Radarcon 2012) in Atlanta, GA, in May 2012 and the tutorial entitled "RF and Digital Components for Radar" that was held at the same conference with my colleague, IEEE Fellow Prof. Maria Greco.

According to the main theme of Radarcon 2012, "Ubiquitous Radar: Opportunity, Needs, and Solutions for Innovative Radar," the plenary talk and the tutorial presented recent advances in silicon technologies, integrated hardware-software architectures, and radar signal processing techniques enabling the realization of highly integrated ubiquitous radars with low cost, compact size, and low power consumption.

Since then, accepting the invitation of Mark Walsh from Artech, the team of authors for the book has been enlarged to benefit of the experience acquired in several R&D laboratories in academia and industry: the signal processing and remote sensing lab, the RF and microwave IC lab, and the electronic systems lab at the University of Pisa (Professors Maria Greco, Bruno Neri, and Sergio Saponara, respectively) and the Radio Frequency Advanced Design Center (RF-ADC), a joint research center of University of Catania and STMicroelectronics (Dr. Egidio Ragonese and Prof. Giuseppe Palmisano).

As discussed in the book, radar has some key characteristics that make it a unique contactless sensor solution with respect to other competing technologies for large market applications (laser, lidar, visible or infrared cameras, ultrasound, capacitive sensors, etc.). Radar is a sensor capable of operating in all weather conditions and gives to the user a multitude of information and data about the sensed target: if there is a target or not, its distance, its relative speed, and its direction of arrival. Moreover, radar works with nonionizing radiation and hence can be used for vital sign contactless monitoring in biomedical applications, and often it can operate in harsh environmental conditions (extreme temperatures, humidity, radiation levels, etc.).

To reach this goal, the approach to radar design and the performance required of radar should be different than traditional approaches.

The development of a conventional radar technique was mainly pushed by military applications during the Second World War with high-power, large-sized, and long-distance radars. As well, when adopted for civil applications, radar is typically a complex system where the design driver is the optimization of its performance rather than the minimization of its cost, size, weight, and power consumption.

As a consequence, radar is a niche market technology for professional applications at the state of the art. To maximize the achievable performance, a "traditional" radar is typically implemented as the interconnection of a set of multiple circuit boards, using high-performance devices, often custom-designed, by mixing different technologies since for each radar subsystem the most suitable technology has to be used.

In this book, we discuss how a different approach can be followed for highly integrated low-power radar design: similarly to other large market electronic design solutions, standard and commercial devices/technologies can be used for radar. The trend should be toward using the same technology for all radar subsystems with the intent of a system-on-chip or system-in-package integration. The aim of the radar designer should not be to maximize radar performance but to optimize the trade-off between the achievable performance and the relevant cost, size, and power consumption.

Obviously, an integrated low-power radar aims at a system-level performance that is quite different from that of conventional radars. Indeed, the maximum operating distance can range from 1m (e.g., in the case of contactless measure of heart rate or breath rate) to maximum 100m to 200m (long-range automotive radar for automatic cruise control). The transmitted power is often below 20 dBm (100 mW). The cross section of the targets (that can be small yachts, cars, pedestrians, bicycles, or parts of the body in the case of biomedical applications) can be from some cm^2 to a few m^2. The size of the radar should be less than 10 cm per side so that the radar can easily be mounted behind the front cover of a car in automotive applications or can be a handheld device for other applications. Short wavelengths should be adopted to ease the miniaturization process.

As analyzed in the book at the algorithmic level, the reader is not required to have state-of-the-art radar signal processing know-how. Rather, selection and optimization of known pulsed or continuous-wave techniques for application scenarios is required where the final aim is not the maximum achievable performance but providing a service with low cost, low power consumption, and low size.

Implementation examples proposed in the book will show that what is discussed is not limited to theory but practical implementations are already possible for a lot of medium- or large-volume market applications: automotive short-range radars for car parking, side-crash warning, collision warning, blind-spot detection and stop and go control in urban scenarios; automotive long-range radar for adaptive cruise control; short-range radars for contactless heart and pulmonary monitoring in e-health applications; and networks of low-cost radars in a harbor for small yachts.

In conclusion, the authors hope that the content of the eight chapters of the book will provide the readers with interesting samples of present research and implementation activities on emerging ubiquitous radar applications.

Prof. Sergio Saponara
May 2014

Acknowledgments

The authors acknowledge the contribution of Riccardo Massini and of the research group of Professors Fabrizio Berizzi and Enzo Dalle Mese, all from University of Pisa, Italy, for the work on the radar for traffic harbor control discussed in Chapter 6.

For the 24- to 77-GHz radar implementation examples discussed in Chapter 7, the authors acknowledge the contribution of the millimeter-wave research team of the RF-ADC, especially of Dr. Angelo Scuderi (now with STMicroelectronics, Catania), Dr. Giuseppina Sapone (now with Infineon Technologies, Munich), and Dr. Vittorio Giammello (now with STMicroelectronics, Catania).

The authors are very grateful to the reviewers who provided valuable comments and suggestions to improve the quality of the book.

They also acknowledge the Artech team, particularly Mark Walsh and Aileen Storry, for their careful support.

1

Scenarios, Applications, and Requirements for Highly Integrated Low-Power Radar

The basis of radio detection and ranging (radar) theory and techniques dates back to the beginning of the twentieth century. The development of such technology was mainly pushed by military applications during the Second World War with high-power (up to several kilowatts and more), large-size, and long-distance radars.

The radar technique has been adopted also for civil applications (e.g., long-range radar (LRR) for obstacle detection on-board airplanes and ships, traffic control in airports, long-distance measurements, and remote sensing). These examples of civil radar applications still refer to cumbersome systems, where the design drivers are typically the optimization of radar performances while its cost, size, weight, and power consumption are not the key issues.

As a matter of fact traditional radar technology is not a "consumer" technology with a large volume market where devices are available at low cost and low power consumption and entire systems are embedded in a single electronic board or chip. Rather, radar technology is a niche market technology for professional applications.

At the state of the art, a radar is typically realized as the interconnection of a set of multiple circuit boards [1], each dedicated to a specific subsystem: RF or microwave front-end with antenna, passives, and switch, often with different boards for transmitter and receiver; analog to digital (A/D) and digital to analog (D/A) converter boards; baseband signal processing board; power supply board; and user interface board.

To optimize the achievable performances, a radar typically relies on high-performance devices, often custom designed, mixing different implementation technologies since for each radar subsystem the most suited technology is used.

This approach is completely different from large-market electronic designs, where standard and commercial devices are used and the trend is using the same technology for all subsystems, and where the aim is optimizing the trade-off between achievable performances and the relevant cost, size, and power.

For example, analog front-end electronics for traditional radar relies on high-performance III-V devices such as high electron mobility transistors (HEMT) in GaAs or InP materials [2, 3]. These technologies allow for high-frequency applications, maximizing circuit metrics such as noise figure, amplifier gain, and linearity transmitter power; as a consequence, high-end performances at the device/circuit level allow for optimal radar performance at the system level (target distance, sensitivity, false alarm rate, and so on). However the cost and design time for III-V–based technologies is much higher than standard silicon transistor solutions such as complementary metal-oxide semiconductor (CMOS) and can be sustained only for professional markets (e.g., defense).

Similarly, the baseband digital signal processing of radar high-performance computing platforms such as GPUs or application-specific integrated circuits (ASIC) are often adopted at the state of the art; unfortunately, their cost and power consumption is much higher than embedded platforms used in consumer electronics.

Recent advances in silicon technologies and electronic design methods make possible the realization of highly integrated radars with low cost, compact size, and low power consumption with achievable performances that can be enough for a large set of applications [4–45] such as automotive short-range radars (SRRs) for car parking, side-crash warning, collision warning, blind-spot detection and "stop and go" control in urban scenarios; automotive LRR for adaptive cruise control (ACC); SRR for contactless heart and pulmonary monitoring in e-health applications; SRR for vital signs detection in case of natural disasters or war scenarios; distance measurements in industrial automation; millimeter-wave body scanner for security (e.g., in airports and banks); networks of low-cost radars for traffic control in railway crossing or in a small harbor for small yachts.

So, the aim of the book is presenting emerging technologies, new circuits solutions, and platform implementations (Chapters 2 and 3) that can enable the widespread adoption of radar for new civil and defense applications. Some example applications for contactless vital signs detection (heart rate and breath rate), harbor traffic control, and automotive driver assistance will be discussed as case studies in Chapters 6 and 7.

As will be further discussed in the book (Chapters 4 and 5) the selection and optimization of known pulsed or continuous-wave techniques for application scenarios at the alogrithmic level is required where the final aim is not achieving the maximum possible performances but reaching sufficient performances to provide a service with low cost, low power consumption, and small size.

The question of whether there is a large volume market for radars in the future arises. Similarly, is important to highlight which are the key characteristics that make radar a unique solution with respect to other competing technologies.

Some answers can be derived by analyzing one of the possible large-volume target applications of low-power radar: the automotive scenario.

According to forecast market analysis [46, 47], in 2014, 7% of all vehicles sold worldwide, particularly in Europe, Japan, and the United States, will be equipped with radar systems with a market value of several billions of U.S. dollars. In a few years, this value can grow to the order of 10% to 20% of new cars with at least two different radar devices (one SRR covering the rear and one LRR covering the front) mounted on each car. This market forecast can become real if the cost for the customer of an automotive radar will be less than $1,000; thus, radars will be mounted not only on top models of premiums car brands but also for medium-range cars. The size of the radar should be less than 10 cm for each side so that a radar can be easily mounted behind the front cover of a car.

As will be discussed in subsequent chapters, for such a radar, the modulation schemes can be pulsed ultrawideband (UWB) for SRR applications within 10m, and frequency modulated continuous wave (FMCW) for mid- and long-range applications as high as several hundreds of meters.

SRRs operate below 10 GHz or around 24 GHz, while LRRs operate at higher frequencies, such as 77 to 81 GHz, where there are less strict limitations in terms of output power spectral density (PSD) [46, 48].

Figure 1.1 reports that for European Telecommunications Standards Institute (ETSI) and Federal Communications Commission (FCC) standards, the equivalent isotropically radiated power (EIRP) spectral density expressed in dBm/MHz [46, 47]. Note that at 77 GHz there is a path loss of roughly 150 dB at 100m distance, and hence high-gain antennas and high-sensitivity receivers are required for LRR applications.

The output transmitted power of integrated radars typically amounts to few dBm. For power levels on the order of watts, or higher, off-chip power amplifiers are needed.

Figure 1.2 shows the main characteristics of SRR and LRR automotive systems: modulation used (Mod.), center frequency (Freq.) and bandwidth (BW), antenna aperture angle (θ), ranging distance (Range) and resolution

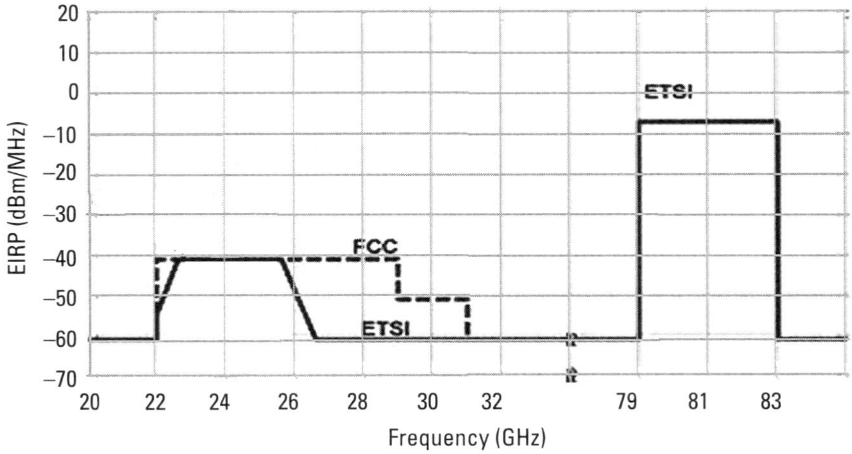

Figure 1.1 EIRP power spectral density in dBm/MHz in ETSI and FCC.

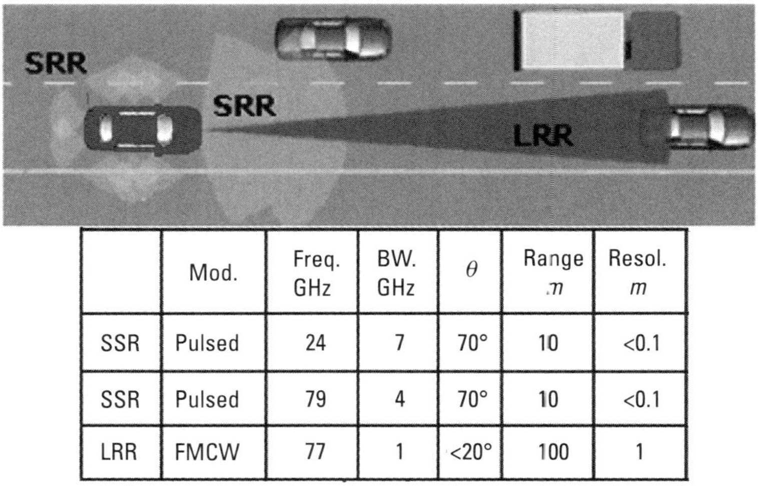

	Mod.	Freq. GHz	BW. GHz	θ	Range *m*	Resol. *m*
SSR	Pulsed	24	7	70°	10	<0.1
SSR	Pulsed	79	4	70°	10	<0.1
LRR	FMCW	77	1	<20°	100	1

Figure 1.2 Automotive radar specifications.

(Resol.), peak power, and PSD. The values of peak power and PSD refer to the limits adopted in Germany [46].

The success of such applications in large-volume markets requires at least the single board integration of mm-wave radar transceiver integrated circuit (IC) and the baseband digital signal processor IC.

The characteristics highlighted for automotive radar can be applied also for other ubiquitous low-power applications. Hence for the target of this book:

- The maximum distance to reach can be less than 1m (e.g., in case of contactless measure of heart rate or breath rate) to a maximum of 100 to 200m (long range automotive radar for automatic cruise control);
- The transmitted power is below 20 dB, often in the range of 10 to 15 dBm;
- The cross section of the targets (small yachts, cars, pedestrians, bicycles, etc.) can be from some cm^2 to few m^2;
- Multiple channels can be useful for diversity channel gain or to extract angle information;
- Short wavelength should be adopted to ease the miniaturization process (e.g., the wavelength is only 2.9 mm at 77 GHz).

At wavelengths of few millimeters, there is potential for high miniaturization, even for antenna integration.

The trend is designing radar system in the following spectrum portions (see Figures 1.3 and 1.4):

- 77–81 GHz suited for LRR and SRR;
- 60 GHz reserved for short-range radio.

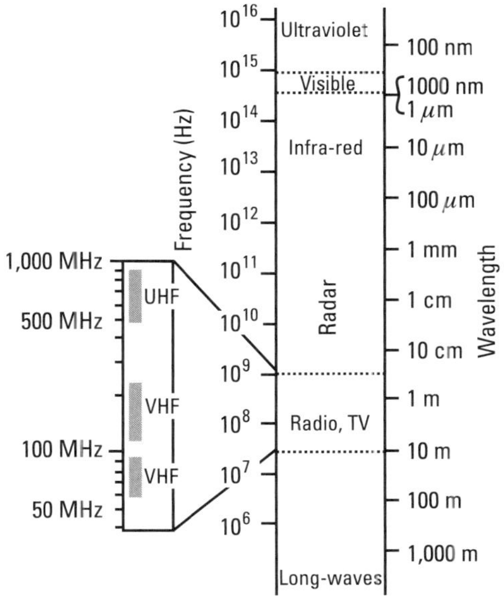

Figure 1.3 Typical radar frequencies within the electromagnetic spectrum.

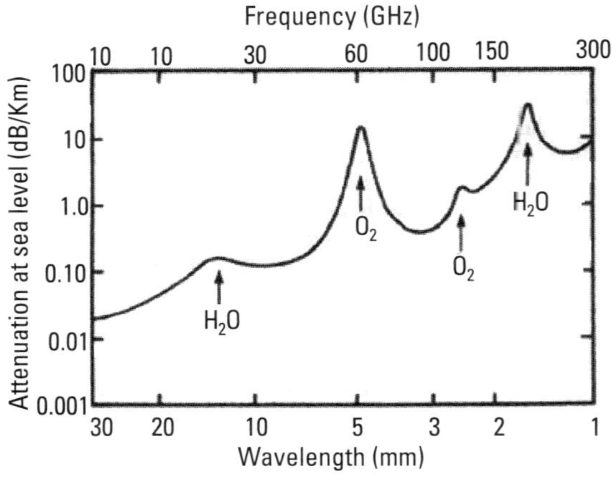

Figure 1.4 Opportunities of radar at mm-waves (60 GHz and 77–81 GHz).

Today, thanks to technology scaling, silicon (Si)-based technologies are offering good characteristics at microwaves and mm-waves. Due to high attenuation, the 60-GHz band (V band), available worldwide for free, is reserved for short communication. At 77–81 GHz (W band), the attenuation is lower; hence, a higher distance can be reached, and there are good opportunities for both LRR and SRR in mm-wave.

It is worth noting that, with respect to other technologies that can be used for contactless sensing, a radar has several advantages:

- A radar can operate in all weather conditions, in bad light conditions, during night and day, while, for example, camera-based systems for obstacle detection have poor performance in bad light conditions or must use a complex multispectral array of sensors (a camera operating in the visible spectrum, plus a camera operating in the near infrared spectrum, and finally a camera operating in the far infrared spectrum portion).
- A radar allows for contactless sensing and no separate line-of-sight sensing, unlike optical or photonic techniques that need a line-of-sight coupling.
- A radar is a nonionizing radiation and hence can be used also for biomedical applications, such as heart rate and breath rate contactless measurement.
- A radar has ground-penetrating capabilities.
- A radar allows for multiparameter sensing, since it can detect whether or not there is a target and can reveal the distance, the relative speed, and the direction of arrival (many sensors give only one form of information).

In this book, the state of the art of design techniques, implementing platforms, and applications of highly integrated low-power radar will be described. Particularly, Chapter 2, "Radar Integration Levels, Technology Trends, and Transceivers," will discuss the possible radar integration levels; the evolution in semiconductor technologies, active devices (e.g., transistors), and passive devices (e.g., antennas waveguide, filters); and integrated transceivers (high-frequency analog radar front-end at both receiver and transmitter sides) toward a deep miniaturization of radar.

Chapter 3, "Hardware-Software Implementing Platforms for Radar Digital Signal Processing," deals with implementing platforms and performance metrics for radar signal processing. A hardware-software architecture for a cost-effective radar baseband digital signal processing (DSP) will be presented after analyzing and comparing different competing solutions such as DSP processors, graphics processing unit (GPU), multicore general purpose processor (GPP), ASICs, and field programmable gate array (FPGAs).

Chapter 4, "Radar for E-Health Applications: Signal Processing Perspective," provides an algorithmic and system view of an integrated radar but for vital signs monitoring applications, particularly the contactless monitoring of heart rate and breath rate.

Chapter 5, "Radar for Automotive Applications: Signal Processing Perspective," gives a system and algorithmic view of radar sensors for automotive systems, their characteristics, and their functions. The signal processing chain for a single 77-GHz FMCW radar sensor will be analyzed in detail: general scheme, equations for range and frequency estimation, constant false alarm rate (CFAR) processing, directions on arrival estimation, and target tracking.

Chapter 6, "Low-Power Radar Front End for E-Health and Harbor Surveillance: Implementation Examples," presents some circuital solutions for two application case studies of a low power integrated radar device. The first is a wearable radar—noninvasive continuous hearth monitoring (NIHM)—for vital function monitoring based on a pulsed radar device integrated in a single-chip in standard CMOS silicon technology.

The second is a frequency modulated continuous wave (FMCW) radar for harbor surveillance integrated at board level, which is realized using microwave integrated circuit (MIC) technology.

Chapter 7, "Automotive Radar IC Design: 24-GHz UWB and 77-GHz FMCW Implementation Examples," presents integrated circuits solutions for UWB 24-GHz radar for short-range automotive applications and 77-GHz FMCW radar for long range applications. All the circuits presented in this chapter are integrated in a silicon-germanium (SiGe) BiCMOS technology.

Finally, conclusions are drawn in Chapter 8.

References

[1] Skolnik, M., *Radar Handbook*, 3d Ed, New York: McGraw Hill, 2008.

[2] Cooke, M., "Technology Focus: Indium Phosphide Semiconductor," *Today Compounds & Advanced Silicon*, Vol. 1, No. 3, 2006, pp. 28–31.

[3] Kang, Dong Min, et al., "A 77-GHz Automotive Radar MMIC Chip Set Fabricated by a 0.15-μm MHEMT Technology," *IEEE IMWS*, pp. 2111–2114.

[4] Lee, J., Y.-A. Li, M.-H. Hung, and S.-J. Huang, "A Fully-Integrated 77-GHz FMCW Radar Transceiver in 65-nm CMOS Technology," *IEEE J. Solid State Circuits*, Vol. 45, No. 12, 2010, pp. 2746–2756.

[5] Li, M., R. Evans, E. Skafidas, and B. Moran, "Radar-on-a-Chip (ROACH)," *IEEE Radar Conference*, 2010, pp. 1224–1228.

[6] Jahn, M., A. Stelzer, and A. Hamidipour, "Highly Integrated 79, 94, and 120-GHz SiGe Radar Frontends," *IEEE MTT-S Inter. Microwave Symp.*, 2010, pp. 1324–1327.

[7] Mitomo, T., et al., "A 77-GHz 90-nm CMOS Transceiver for FMCW Radar Applications," *IEEE J. Solid State Circuits*, Vol. 45, No. 4, pp. 928–937, 2010.

[8] Ragonese, E., et al., "A Fully Integrated 24GHz UWB Radar Sensor for Automotive Applications," *IEEE ISSCC 2009*, pp. 306–307.

[9] Moquillon, L., et al., "Low-Cost Fully Integrated BiCMOS Transceiver for Pulsed 24-GHz Automotive Radar Sensors," *IEEE CICC 2008*, pp. 475–478.

[10] Winker, V., et al., "79GHz Automotive Short Range Radar Sensor Based on Single-Chip SiGe-Transceivers," *EuRAD 2008*, pp. 459–462.

[11] Richter, M. D., et al., "77 GHz Automotive Digital Beamforming Radar with SiGe Chipset," *German Microwave Conf. 2010*, pp. 210–213.

[12] Jain, V., et al., "A Single-Chip Dual-Band 22-to-29GHz/77-to-81GHz BiCMOS Transceiver for Automotive Radars," *IEEE ISSCC 2009*, pp. 308–309, 309a.

[13] Nagasaku, T., et al., "77GHz Low-Cost Single-Chip Radar Sensor for Automotive Ground Speed Detection," *IEEE CISC'08*, pp. 1–4.

[14] Huyn, E., et al., "Design and Implementation of Automotive 77GHz FMCW Radar System Based on DSP and FPGA," *IEEE ICCE 2011*, pp. 517–518.

[15] Kim, Sang-Dong, et al., "Design and Implementation of a Full-Digital Pulse-Doppler Radar System for Automotive Applications," *IEEE ICCE 2011*, pp. 563–564.

[16] Zito, D., D. Pepe, B. Neri, and D. De Rossi, et al., "Wearable System-on-a-Chip UWB Radar for Health Care and Its Application to the Safety Improvement of Emergency Operators," *IEEE EMBS 2007*, pp. 2651–2654.

[17] Li, Changzhi, Xiaogang Yu, Chien-Ming Lee, and Dong Li, et al., "High-Sensitivity Software-Configurable 5.8-GHz Radar Sensor Receiver Chip in 0.13-μm CMOS for Noncontact Vital Sign Detection," *IEEE Trans. Microwave Theory and Techniques*, Vol. 58, No. 5, 2010, pp. 1410–1419.

[18] Li, Changzhi, Xiaogang Yu, Dong Li, and Lixin Ran, et al., "Software Configurable 5.8-GHz Radar Sensor Receiver Chip in 0.13 μm CMOS for Non-Contact Vital Sign Detection," *IEEE RFIC 2009*, pp. 97–100.

[19] Yoon, Yeo-Sun, et al., "MVDR Beamforming for Through-the-Wall Radar Imaging," *IEEE Aerospace and Electronic Systems*, Vol. 47, No. 1, 2011, pp. 347–366.

[20] Dallinger, A., et al., "Coherent Millimeter-Wave Imaging for Security Applications," *IEEE Radar Conference*, 2007, pp. 28–31.

[21] Zito, D., D. Pepe, B. Neri, and D. De Rossi, et al., "Wearable System-on-a-Chip Pulse Radar Sensors for the Health Care: System Overview," *IEEE AINAW*, 2007, pp. 766–769.

[22] Montusclat, S., F. Gianesello, and D. Gloria, "Silicon Integrated Antenna Substrate Benchmarking for MMW Wireless Applications in Advanced CMOS Technologies," *EuCAP*, 2006, p. 680.

[23] Fishler, E., et al, "MIMO Radar: An Idea Whose Time Has Come," *IEEE Radar Conference*, 2004, pp. 71–78.

[24] Gogineni, S., et al., "Monopulse MIMO Radar for Target Tracking," *IEEE Trans. Aerospace and Electronic Systems*, Vol. 47, No. 1, 2011, pp. 755–768.

[25] Pratt, G. T., "Subspace Optimization in Centralized Noncoherent MIMO Radar," *IEEE Trans. on Aerospace and Electr. Syst.*, Vol. 47, No. 2, 2011, pp. 1230–1240.

[26] Grossi, E., et al., "Robust Waveform Design for MIMO Radars," *IEEE Trans. on Signal Processing*, 2011.

[27] Maoz, B., L. Reynolds, and A. Oki, "FM-CW Radar on a Single GaAs/AlGaAs HBT MMIC Chip," *Microwave and Millimeter-Wave Monolithic Circuits Symp.*, 1991, pp. 3–6.

[28] Hafez, et al., "12.5 nm Base Pseudomorphic Heterojunction Bipolar Transistors Achieving f_T = 710 GHz and f_{MAX} = 340 GHz," *Appl. Phys. Lett.*, Vol. 87, 2005.

[29] Komijani, A., et al., "A 24-GHz, 14.5-dBm Fully-Integrated Power Amplifier in 0.18 μm CMOS," *IEEE J. Solid-State Circuits*, Vol. 40, No.. 9, Sept. 2005, pp. 1901–1908.

[30] La Rocca, T., et al., "60 GHz CMOS Amplifiers Using Transformer-Coupling and Artificial Dielectric Differential Transmission Lines for Compact Design," *IEEE J. Solid-State Circuits*, Vol. 44, No. 5, May 2009, pp. 1425–1435.

[31] Yao, T., et al., "Algorithmic Design of CMOS LNAs and PAs for 60-GHz Radio," *IEEE J. Solid-State Circuits*, Vol. 42, No. 5, 2007, pp. 1044–1057.

[32] Suzuki, T., et al., "60 and 77 GHz Power Amplifiers in Standard 90 nm CMOS," *IEEE ISSCC*, Feb. 2008, pp. 562–563.

[33] Seo, M., et al., "A 1.1 V 150 GHz Amplifier with 8 dB Gain and 6 dBm Saturated Output Power in Standard Digital 65 nm CMOS Using Dummy Pre-Filled Microstrip Lines," *IEEE ISSCC 2009*, pp. 484–485.

[34] Shaeffer, D., et al., "A 1.5-V, 1.5-GHz CMOS Low Noise Amplifier," *IEEE J. Solid-State Circuits*, Vol. 32, No. 5, May 1997, pp. 745–759.

[35] Ismail, A., et al., "A 3–10-GHz Low-Noise Amplifier with Wideband LC-Ladder Matching Network," *IEEE J. Solid-State Circuits*, Vol. 39, No. 12, Dec. 2004, pp. 2269–2277.

[36] Adabi, E., et al., "30 GHz CMOS Low Noise Amplifier," in *RFIC Symp. Dig.*, Jun. 2007, pp. 625–628.

[37] Ashfar, B., et al., "A Robust 24 mW 60 GHz Receiver in 90nm Standard CMOS," *IEEE ISSCC*, 2008, pp. 182–183.

[38] Berenguer, R., et al., "A 43.5 mW 77 GHz Receiver Front-End in 65 nm CMOS Suitable for FMCW Automotive Radar," *IEEE CICC 2010*, pp. 1–4.

[39] Chowdhury, D., et al., "A Single-Chip Highly Linear 2.4 GHz 30 dBm Power Amplifier in 90 nm CMOS," *IEEE ISSCC 2009*, pp. 378–379.

[40] Haldi, P., et al., "A 5.8 GHz 1 V Linear Power Amplifier Using a Novel On-Chip Transformer Power Combiner in Standard 90 nm CMOS," *IEEE J. Solid-State Circuits*, Vol. 43, No. 5, 2008, pp. 1054–1063.

[41] Alam, S. K., "A Novel 2.4 GHz CMOS Variable Gain Low Noise Amplifier Design for Bluetooth and Wireless LAN Applications," *IEEE ICCE'07*, pp.1–2.

[42] Wang, S., and B. Z. Huang, "A High Gain CMOS LNA fro 2.4/5.2 GHz WLAN Applications," *Progress in Electromagnetic Res.*, Vol. 21, 2011, pp. 155–167.

[43] Zito, D., et al., "Feasibility Study of a Low-Cost System-on-a-Chip UWB Pulse Radar on Silicon for the Heart Monitoring," *IEEE IWDDDC*, 2007, pp. 32–36.

[44] Chew, K. W., et al., "Impact of Device Scaling on the 1/f Noise Performance of Deep Sub-Micrometer Thin Gate Oxide CMOS Devices," *Solid State Electronics*, 2006, pp. 1219–1226.

[45] Zito, D., et al., "A 90nm CMOS SoC UWB Pulse Radar for Respiratory Rate Monitoring," *IEEE ISSCC*, Feb. 2011, pp. 40–41.

[46] Goppelt, M., H.-L. Blocher, and W. Menze, "Automotive Radar—Investigation of Mutual Interference Mechanisms," *Adv. Radio Sci.*, Vol. 8, 2010, pp. 55–60.

[47] Hoetzer, D., et al., "Automotive Radar and Vision Systems—Ready for the Mass Volume Market," *Vehicle Dynamics Expo*, Oct. 2008.

[48] Australian Communications Authority, RF Planning Group, "A Review of Automotive Radar Systems—Devices and Regulatory Frameworks," Doc. SP4/01, 2001.

2

Radar Integration Levels, Technology Trends, and Transceivers

2.1 Radar Integration Levels

Different levels of integration are possible for a low-power radar, overcoming state-of-the-art solutions where a radar is typically realized as the interconnection of a set of multiple circuit boards [1], each dedicated to a specific subsystem (RF or microwave front-end with antenna, passives, switch, often with different boards for transmitter and receiver; A/D and D/A board; baseband signal processing board; power supply board; user interface board).

Radar integration can be from single-board level to single-chip level with increasing miniaturization but also increased technology complexity [2, 3] (see Figure 2.1). As a matter of fact, the three following solutions can be addressed, with different trade-offs between performance and implementation cost.

2.1.1 System-on-a-Single-Chip

System-on-a-single-chip (SoC) is where the radar is completely contained in a single chip. This solution can allow for the minimum size and weight but also is the most complex to be realized since a radar requires different types of electronic devices (active and passive devices, analog and digital components, low-noise receivers but also a power transmitter) that usually have different optimal target technology [4–7].

This approach would require huge investments and high nonrecurring engineering costs and hence is suitable only for large-volume productions, as

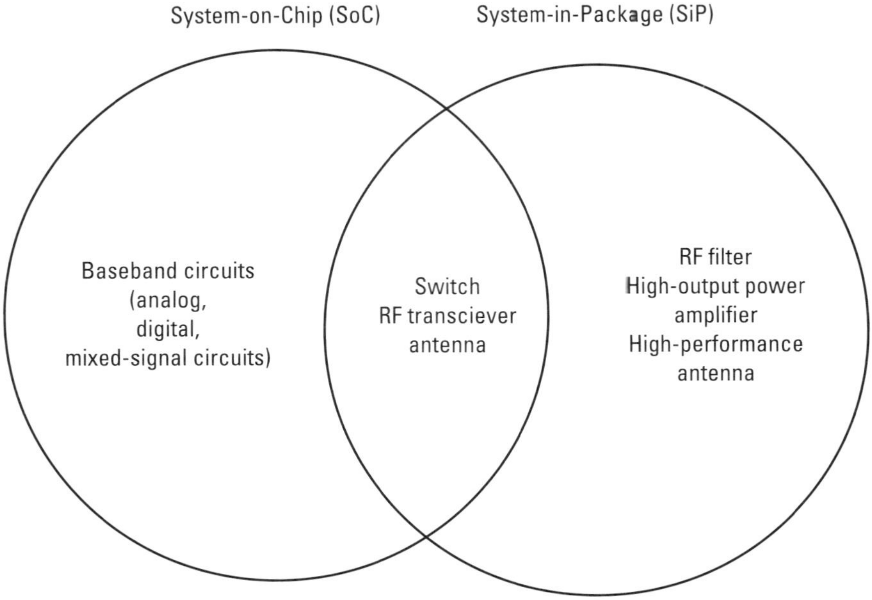

Figure 2.1 Integration domains for system-on-a-single-chip and system-in-a-package radar.

in consumer electronics (e.g., tablets, smart phones, game consoles). However radar is not a large-volume consumer application.

Moreover, having all radar devices on the same chip can cause problems of interference and noise coupling between power and low-noise circuitry, between analog and digital signal domains. The whole radar system performance can be seriously reduced.

2.1.2 System-in-a-Package

System-in-a-package (SiP) is where the radar is realized using multiple chips but embedded in a single package. With respect to the previous SoC solution, with the SiP approach active (analog and digital) devices can be realized as integrated circuits on multiple cores, each of them optimized for a specific radar subsystem and assembled on the same substrate within the single package; passive devices and antenna can be realized exploiting packaging technology.

Different SIP technology options are available or are under research for mm-wave low-power radar or radio applications such as integrated substrate and/or multichip module (MCM), even 3D, or low-temperature cofired ceramic (LTCC) [8–11].

Thanks to the SiP approach, interference and noise coupling problems among the radar subsystems in SoC can be reduced; the achievable performance in terms of miniaturization level is still good compared to state-of-the-art multiboard radar. The production volume required for a SIP solution to be economically convenient is lower than with fully integrated SoC.

Obviously, low size and low power consumption entail low transmit power; this limits the achievable target distance. Therefore, a highly miniaturized radar is de facto a short-range radar. Unless very low power and low antenna gain are required, the SiP approach is a more viable solution for radar than fully integrated SoC approach.

2.1.3 Single-Board Radar

Single-board radar is where the system is realized using multiple integrated circuits mounted on a single board. Compared to SoC and SiP, this approach is the easiest to achieve and the investment required is not so high; hence, it is possible for low-medium production volume as we can expect for radar market.

With a single-board radar approach, all subsystems can be realized using a dedicated integrated circuit with its own package (e.g., CMOS silicon technology can be used for A/D and D/A conversion and baseband digital signal processing), monolithic microwave integrated circuits (MMIC) can be dedicated to the analog RF and microwave parts of the transceivers to reach the desired targets in terms of noise, gain, and linearity.

All the integrated circuits are then assembled on a single board where passive devices can also be added as discrete devices or as integrated device (e.g., using the microstrip/strip-line approach).

Summarizing, increasing integration levels has some advantages and other disadvantages:

- Advantages of highly integrated radar:
  - Component assembly minimized;
  - Increasing reliability;
  - Increased operating lifetime;
  - Small size, small weight, low power consumption;
  - Increased reproducibility and lower cost for large volume production.
- Disadvantages of highly integrated radar:
  - IC design has high nonrecurring costs (CAD tools and foundry cost, design time, and team design cost) and hence cost is minimized only for large volume production;
  - A single technology cannot offer optimal performance for all radar subsystems (e.g., CMOS technology is optimal for baseband DSP, not for antenna or RF circuits);
  - Low transmit power limits possible applications to short range ones.

2.2 Next Steps in Radar Miniaturization

Following the analysis in Section 2.1, today radars with high transmit power and large aperture antenna are realized by assembling multiple electronic boards, each optimized for a specific subsystem (antenna subsystem with feed, reflectors, and scanning modules; traveling wave tubes or Klystron as power amplifier modules; MMIC for TX/RX module; multiple boards for digitization and radar signal processing; user interface and networking) while the next step, for low-power ubiquitous radar, is assembling all subsystems on the same single printed circuit board (PCB) realizing a radar-system-on-a board.

A radar-system-on-a-board will assemble on the same board at least the following components [3]:

- A single chip of a fraction of 1 cm2 (an integrated radar transceiver occupying roughly 9 mm2 will be shown in Chapter 7) integrating the whole transmit (TX) and receive (RX) chains operating in the RF or mm-wave domain. CMOS or SiGe or MMIC in III-V technologies can be used for this purpose.
- Solid-state power amplifier (depending on the transmit power needed).
- A single chip for baseband digital signal processing that can be a digital signal processor (DSP), an FPGA, or a custom IC in CMOS technology supporting the following main tasks: waveform generation, matched-filtered, pulse compression, range/speed ambiguities resolution, CFAR technique, and so on.
- Memory modules, both random access memory (RAM) for data storage while the radar system is on and nonvolatile memories (e.g., flash devices) to store instruction code and data to be reused when the radar is switched off.
- Analog-to-digital and digital-to-analog (ADC/DAC) converters (if not already integrated in the custom IC, CMOS technology).
- Antenna (printed on the PCB board if gain, bandwidth, efficiency are enough).

For a midterm evolution of low-power radar we have to consider that thanks to huge investments from telecom businesses and the consequent submicron technology scaling, CMOS technology is providing good performance for RF and mm-wave low-power transceivers.

The silicon on insulator version, CMOS silicon on insulator (SOI), offers further improved performance at high frequencies and can be suitable for the realization of passive components (inductors, capacitors, even antennas operating at V/W bands if few dB gain are enough) [12–14].

The trend for the midterm future is hence further increasing the miniaturization level with respect to the radar-system-on-a-board integrating the single-chip radar transceiver plus the A/D and D/A converters and part of the DSP chain, such as a fast Fouier transfer (FFT) processor. Only the power amplifier and the antenna will be off-chip.

Next, we address a review of the technology trends for the integration of the most challenging blocks, like antenna and RF transceivers. A detailed analysis of the implementing platforms for digital radar baseband is then detailed in Chapter 3.

2.3 Integrated Antennas

One of the main challenges in the move toward high-miniaturized radar systems is integrating the antenna. At the state of the art radar antennas are typically realized off-chip.

Long-range radar (LRR) automotive applications require antennas with high gain and high directivity, which cannot be realized on-chip. To reach a target distance of several hundred meters, a LRR would require an antenna gain up to 20–25 dB, achieved with a patch or horn or dish antenna in literature [15].

Pulsed radars can use the same antenna in time-division as the TX side and RX side. FMCW radars, such those analyzed from a theoretical and experimental perspective in Chapters 5, 6, and 7, use separate TX and RX antennas.

By using an antenna array, a radar scanning effect can be obtained by realizing beam-forming in the analog domain, through phase shifters, or in the digital domain through digital beam-forming. Unlike beam-forming, which presumes a high correlation between signals either transmitted or received by an array, the multiple-input multiple-output (MIMO) concept exploits the independence between signals at the antenna array elements to improve detection performance.

For radars operating at few GHz frequencies, the wavelength amounts to several cm; hence, it is not convenient to integrate the antenna.

The higher the frequency, the lower the wavelength and hence realizing an integrated antenna, even on chip, becomes feasible: as an example, for 77-GHz radar or 60-GHz radio λ is few mm.

The antenna can be integrated at different levels:

- At board level [printed antenna on printed circuit boards (PCB)]. In the last generation of 77-GHz automotive LRR by Bosch, using INFI-

NEON chipset, as described by J. Hasch et al. in a special issue of *IEEE Transactions on Microwave Theory and Technique* 2012 [4], based on the FMCW principle, four single microstrip patch antennas are combined with parasitic elements to adjust bandwidth and beamwidth. The four antenna elements are tilted by 45 degrees to reduce interference from coming cars. The antenna elements in the Bosch solutions serve as feeds for a further dielectric lens resulting in four narrow beams.

- At package level (e.g., using low-temperature cofired ceramic (LTCC) technology to realize multilayer circuits with integrated passive components, including the antenna). An LTCC-integrated example of a CW-radar antenna plus transceiver for near-field high accuracy measures in industrial scenarios has been proposed by C. Rusch et al. at IEEE EuCAP'11.
- At chip level using MMIC or silicon on insulator technologies. Lots of on-chip antenna designs at 60 GHz for short-range consumer radio have already been proposed in the past years by academia and industry: as an example, a double-slot antenna at 60 GHz by Huei Wang et al. at IEEE SIRF 2010 [16] or complete V-band transmitter with integrated antenna in a special issue in *IEEE Microwave* magazine in 2009 [17].

However there is lot of work still to do to achieve the high antenna gain required by radar systems on a small, easily fabricated chip.

The antenna conceived for 60-GHz consumer radio applications have much less stringent performance than for typical radar systems. The 60-GHz short-range radio applications for consumer applications are typically characterized by low gain antenna with broad beamwidth.

Better performance can be achieved using SOI technology; antennas integrated in SOI devices have been recently proposed in literature for 60 GHz and 77 GHz with a gain still limited to few dB [11, 12].

In SOI technology, the high resistivity of the substrate on which n- and p- metal-oxide-semidconductor field-effect transistor (MOSFET) are created allows dielectric isolation of circuit elements. Therefore, junction capacitances are reduced, increasing maximum operating frequency, and there is a reduced noise coupling between digital and analog parts integrated in the same chip.

The performances of coplanar stripline (CPS), coplanar waveguide (CPW), or antennas in SOI CMOS are improved due to a reduced amount of energy loss in the supporting substrate.

For example, some works by STMicroelectronics (F. Gianesello et al. in IEEE SOI 2010 conference) show the incidence of substrate resistivity on achievable radiation efficiency and gain of an integrated antenna: moving from bulk semiconductor to SOI technology (substrate resistivity of bulk 20 Ω/cm to substrate resistivity of SOI 1000 Ω/cm) the radiation efficiency of a designed antenna grows from a poor level of 6% to roughly 25%. The antenna gain grows from –8 dBi to almost 0 dBi when increasing the substrate resistivity from bulk CMOS to SOI CMOS.

At the University of Pisa, we have recently designed integrated antennas in 65-nm CMOS bulk and 65-nm CMOS SOI technology: an inverted-F antenna, a double-slot antenna, and a bow-tie antenna. The double-slot antenna in 65-nm CMOS SOI technology with coplanar waveguide feed achieves a gain of 4.4 dBi at 60 GHz and has an area occupation of few mm^2. The CMOS bulk technology has lower cost, but the achievable antenna performance is lower. With respect to the double slot, the inverted-F antenna has lower area occupation but lower gain while the bow-tie antenna has large bandwidth.

Figure 2.2 shows the achievable radiation pattern and gain for the double-slot integrated antenna designed at the University of Pisa in CMOS 65-nm SOI technology (HFSS 3D electromagnetic simulator).

Figure 2.3 shows the achievable S11 performance and the input impedance as a function of frequency for the double-slot integrated antenna designed at the University of Pisa in CMOS 65-nm SOI technology (HFSS 3D electromagnetic simulator).

Therefore, integrated antennas are useful only for short-range applications. Their performance can be increased using a special dielectric lens antenna or smart resonator. For example, J. Hash et al. [4] for a new generation of automotive radar proposed on-chip antenna elements based on shorted $\lambda/4$ micro-strip lines, formed by the top and bottom metal layers of the chip back-end. However most of the radiation is dissipated due to conductor and dielectric losses, resulting in a low antenna efficiency (<10%). To improve the antenna performance, quartz glass resonators are positioned above the on-chip

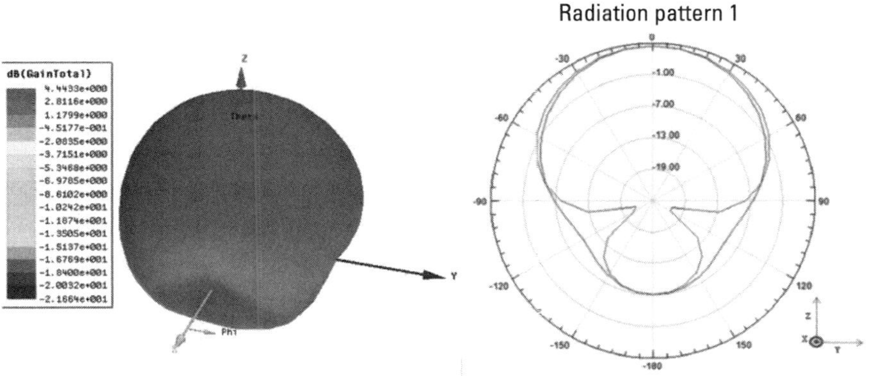

Figure 2.2 Radiation pattern and gain at 60 GHz of a double-slot integrated antenna in CMOS 65-nm SOI-achievable technology.

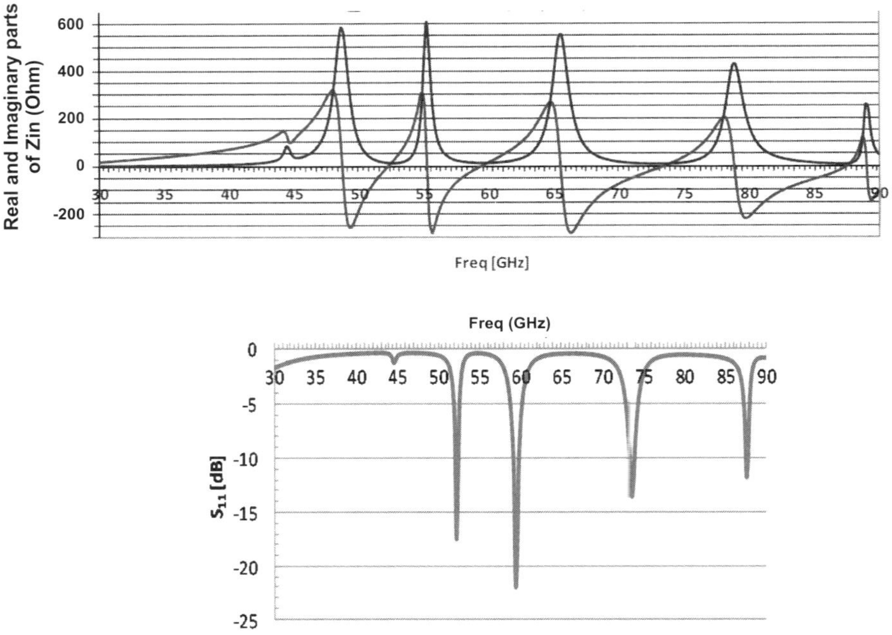

Figure 2.3 S11 performance and input impedance as a function of frequency for a double-slot integrated antenna in CMOS 65-nm SOI technology.

patch antenna elements to improve efficiency and bandwidth. The antennas are spaced at a distance to allow direction of arrival (DOA) estimation of a target or provide separate beams illuminating a dielectric lens.

Table 2.1 summarizes the main data (antenna type, central working frequency, target implementing technology, performance in terms of gain and bandwidth, type of feeder, and input impedance) for integrated antenna designs recently proposed in literature.

2.4 Semiconductor Technology and Devices for Integrated Radar

Different semiconductor technologies are today available, each suited for the optimal design of a specific radar subsystem.

Table 2.2 shows the key characteristics (energy gap, breakdown voltage, carrier mobility, carrier speed saturation, thermal conductivity) of some semiconductor technologies, silicon, and heterojunction devices.

From reported data is clear that III-V high-mobility devices (GaAs, InP, and so on) are the most suited for high performance at high frequencies (electrons rather than hole carrier–based devices).

Table 2.1
Performance of Integrated Antenna Designs Recently Proposed in Literature

Antenna Type	F (GHz)	Tech	Gain	BW(GHz)	Feeder	Imped.
4 array dipole	77	SiGe	2	2	Differential	45Ω
Slot dipole	24	GaAs	2	1.4	CPW	50 Ω
Zig zag	24	CMOS	1.5	N/A	N/A	30 Ω
Aperture coupled patch	60	CMOS	7	7.8	Balanced	100 Ω
Dipole	60	SiGe	2.35	7	CPS	30 Ω
Slot antenna	60	CMOS	10	5	N/A	N/A
Cavity backed folded dipole	60	SiGe	7	18	CPS	50 Ω
Folded dipole	60	SiGe	8	8	CPW	100 Ω
Yagi	60	SiGe	7	9.4	N/A	50 Ω
Spiral	60	CMOS SOI	4.2	15	CPW	50 Ω

Table 2.2
Key Characteristics of Main Semiconductor Materials

Material	Eg	Vbreak	Mobility Electrons	Vsat Electrons	Mobility Holes	Vsat Holes	K
	(eV)	(10^5 V/cm)	(10^3cm^2/V*s)	(10^7 cm/s)	(10^3cm^2/V*s)	(10^7 cm/s)	(W/K*cm)
Si	1,12	3	1,5	1	0,6	0,7	1,5
GaAs	1,42	4	8	0,8	0,4	0,9	0,5
InP	1,35	5	5	0,7	0,2	0,5	0,67
GaN	3,4	30	1,2	1,5	0,05		1,3
InAs	0,36	0,4	25	0,9	0,5	0,5	0,28

Devices realized in GaN technology, due to wide energy-bandgap and breakdown voltage, are the most suited for the high voltage and high power levels required in radar transmitter (e.g., they can allow for vacuum tube replacement in high transmitter power radar).

Si-based devices are suited for large-volume, low-cost applications since they already dominate baseband analog and digital processing for telecommunications, computers, and consumer electronics.

At the device level, the possible choices are as follows:

- Field effect transistors (FETs) as metal oxide semiconductors (MOS) or HEMTs, which are unipolar devices (single-carrier: electrons in HEMT and NMOS, holes in PMOS);
- Heterojunction bipolar transistor (HBT), which is a bipolar devices (holes and electrons).

HBT and FET can act both as digital devices (on/off) or analog devices (current generator controlled by an input current in HBT or current generator controlled by an input voltage in a FET).

However, some performance metrics are different between bipolar and FET devices:

- HBT has an exponential drive characteristic, while for FET it is quadratic; HBT has typically higher transconductance (gain) than FET.
- HBT has a resistive input impedance, while for FET it is capacitive; as consequence, FET has lower power consumption in standby mode.
- FET has lower noise at high frequency, but it suffers of 1/f noise at low frequency.
- HBTs are vertical integrated devices (high current density), while low-power FETs are planar devices (higher integration density and easier scaling).

In highly integrated low-power radar, the vacuum-tube technologies (traveling wave tube, klystron, or magnetron) adopted for high-power high-performance radar transmitters are completely avoided.

Competing technologies for integrated radars are solid state ones:

- MMICs in compound III-V semiconductors such as GaAs, In-P using high electron mobility FET HEMTs;
- For power amplifiers, interest in GaN and SiC is increasing;
- SiGe HBT or bipolar-MOS (BiCMOS) ICs;
- Si CMOS (N- and P- MOS) technologies ICs;
- CMOS SOI ICs.

Table 2.3 summarizes for the competing technologies some key circuit figures of merit:

Table 2.3
Performance of Competing Technologies

	F_T	Gain/NF Ratio	Cost	Power Consumption	Suited for
BJT	High	High	Medium	High	Analog, RF
CMOS	Medium	Medium	Low	Low	Digital
BiCMOS	High	High	Medium	Medium	Analog, RF, mixed-signal
HEMT	Very high	High	High	Medium	mm-wave

- Maximum working frequency represented by Ft, the frequency at which the current gain with short-circuit output is 1;
- The ratio between the achievable gain and the noise figure (NF);
- The achievable power consumption;
- The implementation cost;
- The most suited application domain.

2.5 Trends in IC Radar Design

2.5.1 MIC and MMIC Technology

RF and microwave integrated circuits entirely realized (interconnections and matching networks, active and passive devices) on a single chip (Si, GaAs, or other) are called monolithic microwave integrated circuits (MMIC). They are suited just for large-scale production, in that the unitary cost of the masks and the access to the technological plants necessary for their realization would make the final product too expensive.

III-V MMIC for radio and radar applications have been developed at high frequencies since 1970s and 1980s, thanks to dedicated U.S. funding programs for MMIC technology.

MMIC is now a mature technology, offering for analog circuitry (active and passive components) at microwaves and mm-waves best in class performances in terms of maximum achievable Ft (the frequencies at which the short-circuit current gain is 1), NF, gain of low-noise amplifier (LNA), and gain and saturation power of the power amplifier.

MMIC dominates high-end transceivers from tens of GHz to THz.

Most of MMIC are in GaAs technology (automotive radar front-end at 77-/79-GHz, 60-GHz applications, 94-GHz imaging, Ka-, V-, and W-band radars).

Since 2005, III-V HEMT devices with Ft and Fmax (the frequencies at which the power gain is 1) higher than 700 GHz are available; therefore, III-V devices are allowing for THz domain applications.

However, due to niche market applications, and higher device size, the cost of ICs with III-V technologies is higher than that of silicon technologies. While such cost is affordable in military or space applications, for low-cost and low-power civil radar applications silicon technologies must be used for IC.

An alternative to MMIC for small and medium production scale and for prototype realization and testing is planar monolithic integrated circuits (MICs). In this case, passive elements can be directly realized on the substrate, whereas active ones have to be purchased on the electron device market and bonded on the substrate. The substrate is a "sandwich" made by a dielectric sheet, several hundred microns thick, with two conductive layers (Cu, Au) deposited on the upper and down surfaces. The realization of MICs does not require specific technological plants or expensive processes in that the only necessary technological steps are (1) a photo-lithographic process for conductive layer etching, and (2) a bonding work station for discrete devices mounting on the substrate.

In planar MIC circuits, in addition to single passive or small-signal active devices, more complex devices can be mounted, such as integrated LNA—available on the market in surface-mounting device (SMD) packages or directly on chips—integrated switches, SMD filters, as well as power transistors for output power in the range of tens of Watts. Therefore, MICs are well suited for the realization of complete microwave systems up to 100 GHz in the power range up to tens of Watts. These characteristics make MIC technology eligible for the realization of prototypes, or small- and medium-scale production, which can be the case of low-power, low-cost radar for specific applications. A more detailed description of MIC technology is contained in Chapter 5 together with a case study (implementation and full experimental characterization) consisting of the realization of a radar for specific applications.

Like MMIC, MIC devices suffer the limit of poor digital and mixed-signal integration capability and hence are not suited to low-cost, large-volume, digital-based applications.

2.5.2 Si-Based Technology

Si-based technology dominates electronic industry for baseband (BB) signal processing and intermediate frequency (IF) allowing BB and IF circuitry to be integrated in the same SoC. Since 2000 Si-based technologies (SiGe bipolar or CMOS) are used also for telecom RF ICs (cellular phone transceivers, wireless local area networks (WLANs), Bluetooth, UHF wireless sensors).

Recent technology scaling allows for the use of CMOS, CMOS SOI, or BICMOS also for mm-waves. Si-based mm-wave SoC have been developed in recent years with commercial technologies for automotive radar (24 GHz and now 77/79 GHz) or telecommunication high-bandwidth radio (60-GHz short-range radio).

A rich set of technologies is available from different vendors:

- SiGe BiCMOS 130 nm and 180 nm;
- CMOS or CMOS SOI 130 nm, 90 nm, 65 nm, 45 nm, 32 nm, 28 nm, and 20 nm;
- Full depleted (FD)-SOI at 28 nm and smaller.

For radar applications SiGe BiCMOS 130-nm and 180-nm technology nodes or CMOS 90-nm and 65-nm technology nodes have been already used in literature by academia and industry. The last chapter of this book will present two case studies of radar implementation for 24-GHz and 77-GHz automotive radar in Si-based technologies from STMicroelectronics.

In radar design, HBT are more suited for high-frequency analog circuitry ensuring higher gain and cut-off frequency and lower NF.

MOSFETs are more suited for the baseband DSP due to lower power consumption, easier device scaling, higher integration levels, and lower cost.

MOS technology is the dominating one for processors, memories, and mixed-signal circuitry in the baseband frequency domain.

SiGe bipolar complementary MOS (BiCMOS) allows the cointegration of BJT for high-frequency applications and MOS devices for digital circuits, although at higher cost.

At the state of the art, the SiGe BiCMOS technology, with 130-nm transistors channel length and an Ft of 230 GHz, offers a good trade-off between cost and performance for single-chip mm-wave radar transceivers. Several transceivers at 24, 77, 90, and 120 GHz have been proposed in the literature using SiGe BiCMOS technology.

For the future, for large-volume applications of systems operating at mm-waves (e.g., 60-GHz radio and perhaps automotive radar), the trend will be using CMOS also for mm-wave circuits.

As an effect of device scaling, an Ft higher than 150–200 GHz can be obtained today with MOS devices. Using a rule of thumb that we can safely work at one-third of Ft this means that frequencies up to 60–70 GHz can become also a domain of MOS devices.

It should be noted that technology evolution allows for higher Ft at a given current or the same Ft for lower current: this reduces power consumption, power supply, and thermal issues reducing size and cost and increasing reliability in harsh environments

A mm-wave transceiver in scaled CMOS technology, as baseband DSP, entails a lower area, higher integration, and lower cost for large-volume markets but also lower performance versus 130-nm BiCMOS SiGe technology.

Technology benchmarks realized by A. Scavennec et al., in a special issue of *IEEE Microwave* magazine in 2009 [18], by comparing the phase noise and the output power of a voltage controller oscillator in the range 60–400 GHz found in different technologies (CMOS, HBT SiGe, HBT GaAs) show that CMOS allows for competing performance versus SiGe and GaAs HBT up to 60–70 GHz.

A similar analysis has been carried out in the RF laboratories of the University of Pisa. Considering as a benchmark a low noise amplifier and its achievable performance in terms of gain (dB) and NF (dB) from Figure 2.4, we can

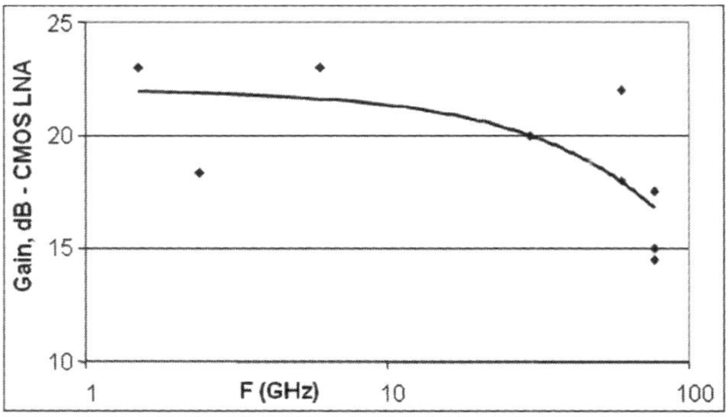

Figure 2.4 Achievable LNA gain in CMOS technology as a function of frequency (1- to 100-GHz range).

see that state-of-the-art designs up to 10–20 GHz in have good performances: gain higher than 20 dB. At higher frequencies, the performances start decreasing. Around 77 GHz (W-band), acceptable but nonoptimal performance is achieved (gain lower than 20 dB).

State-of-the-art designs up to 10–20 GHz in CMOS technology have optimal NF performances, lower than 4 dB (see Figure 2.5). At higher frequencies the performances start decreasing. Around 77 GHz (W band), acceptable but nonoptimal performance is achieved (NF higher than 4 dB).

At the University of Pisa, we also analyzed the achievable saturation power for a transmitting power amplifier (PA) realized at mm-wave in CMOS technology (see Figure 2.6). At frequencies of a few GHz, an integrated PA up to 1W peak power is possible, but the peak power of integrated PA decreases with frequency.

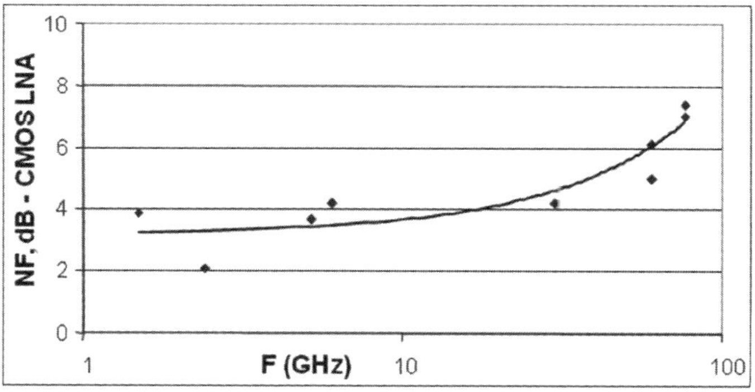

Figure 2.5 Achievable LNA noise figure in CMOS technology as a function of frequency (1- to 100-GHz range).

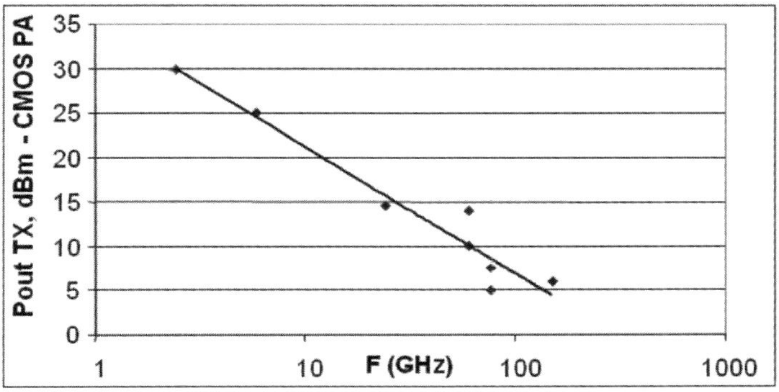

Figure 2.6 Achievable saturate power for the PA (dBm) in CMOS technology as a function of frequency (1- to 100-GHz range).

At high frequencies (77 GHz or higher, W band), the peak power is less than 10 dBm (10 mW). Therefore, only short-range applications are possible; external off-chip PAs are needed.

The real issue still limiting a full CMOS realization of an automotive LRR application is the on-chip mm-wave power amplifier. This a big issue considering that from a radar equation the range capability heavily depends on transmitted power levels.

From the analysis of Scavennec et al. [18] the maximum saturation power at the transmitter side achievable with MOS technology is on the order of 10 dBm (10 mW) at 60–70 GHz, while with SiGe this value grows up to 20 dBm (100 mW) and for the HBT GaAs it is even higher. This is why today's chipsets for LRR automotive applications for the mm-wave part are mainly based on SiGe technology (HBT or BiCMOS) [4, 19]. This will be also the case of the implementation case studies from industry presented in Chapter 7 of this book.

Table 2.4 summarizes the characteristics (process name, technology node, value of Ft and Fmax for a NPN-bipolar transistor, and available metal layers for interconnection) of HBT or BiCMOS SiGe technology adopted by the main Silicon Foundries (IBM, TowerJazz, Freescale, STMicroelectronics, NXP).

As an example, for the IHP foundry, the SG13S SiGe technology is a high-performance 0.13-μm BiCMOS with npn- hetrojunction bipolar transistors (HBTs) up to f_T/f_{max}= 250/300 GHz, with 3.3 V I/O CMOS and 1.2 V logic CMOS. The SG13G2 version has the same device portfolio but bipolar performance with f_T/f_{max} = 300/500 GHz.

2.6 Radar Transceivers

In the previous paragraphs, the focus has been mainly on the analog subsystems of a radar and the relevant RF or millimeter-wave components [low noise am-

Table 2.4
Performance of SiGe Technologies (HBT and BiCMOS)

Parameter	Unit	IBM	TowerJazz	IHP	STM	Freescale	NXP
Process		8HP	SBC18H3	SG13	B9CMW	HiP6MW	QUBiC4Xi
Node	Nm	130	180	130	130	180	250
NPN Ft	GHz	200	240	250	230	185	216
NON Fmax	GHz	265	260	300	290	260	177
NPN gain		450	-	900	950	425	-
NPN BVce0	V	1,7	1,2	1,7	1,5	2	1,4
Metal Layers	N.	4Cu, 2 Al	Al	6 Al	6 Cu	5 Cu	6 Al
MIM capacitor/ Varactor		Y/Y	Y/Y	Y/N	Y/Y	Y/Y	Y/Y

plifier (LNA) power amplifier, oscillator, and antenna]. However, radar is no more a simple analog system, but is becoming a mixed-signal system.

Some key blocks remain in the analog domain such as LNA, antenna switch in case the same antenna is multiplexed in time domain between transmitter and receiver (e.g. in pulsed radars), power amplifier, mixer, oscillators (phase locked loop, voltage controlled oscillator, quartz oscillator, phase detectors, phase shifter, frequency dividers), some IF or baseband analog blocks such as adaptive gain control (AGC) amplifier, baseband amplifiers, and filters or integrators.

However, in literature from academia but also in industry lots of ADCs with pipeline, or time-interleaved architectures that allow for power-efficient conversion with high effective number of bits (ENOB) and high sampling frequency performance are already available [20–24].

Therefore, the radar signal can be digitized directly at IF minimizing the daunting tasks to be carried out in the analog domain at radio frequency. The mixer and the local oscillator (LO) still remain in the analog domain but all the signal processing moves in the digital domain as reported in Figure 2.7.

The requirements on the ADC for a good radar design are as follows:

- ADC operating at intermediate frequency and not only at baseband frequency: sampling rates up to tens, or even hundreds, of MS/s.
- The number of ADC channels depends on the system architecture (1 or 2 for I-Q signal components for each of the K radar channels).
- The bit resolution is typically higher than 10 bit (e.g., a nominal 14–16 bit is required for 12–14 bit ENOB—at least 70 dB dynamic range).
- Specification on nonlinearity and aperture uncertainty of the ADC depends, together with ENOB bits N, also on the required signal-to-noise ratio (SNR).

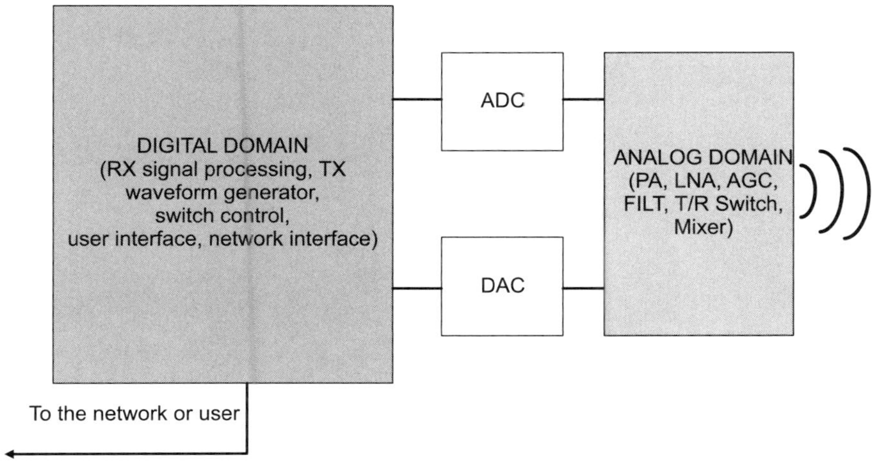

Figure 2.7 Radar as a mixed analog-digital system.

So in the digital or mixed domain we found the following blocks:

- A/D and D/A converters;
- FFT and its inverse (IFFT) processors;
- Timed-domain and frequency-domain digital filters;
- Pulse compression (PC);
- Digital down- and up-conversion (DDC and DUC);
- Beam-forming and more in general waveform generation by direct digital synthesis (DDS) exploiting numerical controlled oscillator (NCO), digital delay locked loop (DLL), or digital clock manager (DCM);
- CFAR;
- Space time adaptive processing (STAP);
- DOA estimation;
- All data processing tasks related to tracking, detection, classification, and data analysis/mining;
- All control and interface tasks since the radar is often a node of a more complex networked system.

It should be noted that the block diagram vision of the radar in Figure 2.7 is generic and is still valid for different radar architectures that have been proposed in literature and that will be analyzed and discussed in the next chapters such as the following:

- Pulsed or continuous-wave (CW) radar;
- Homodyne or heterodyne receiver or correlator-type receiver;
- Down-conversion in the analog domain (the ADC works in baseband) or digital domain (the ADC works at intermediate frequency and an NCO is also used);
- Single antenna or multiple antennas.

With reference to the radar as a mixed-signal system, the analysis of the digital implementing platform is further detailed in Chapter 3.

References

[1] Skolnik, M., *Radar Handbook*, 3rd Ed., New York, McGraw Hill, 2008.

[2] Bhor, M., "The New Era of Scaling in an SoC World," *Proc. 2009 IEEE International Solid State Circuits Conference (ISSCC)*, San Francisco, CA, Feb. 8–12, 2009, pp. 23–28.

[3] Neri, B., and S. Saponara, "Advances in Technologies, Architectures and Applications of Highly Integrated Low-Power Radars," *IEEE Aerospace and Electronic Systems*, Vol. 27, No. 1, 2012, pp. 25–36.

[4] Hasch, J., et al., "Millimeter-Wave Technology for Automotive Radar Sensors in the 77 GHz Frequency Band," *IEEE Transactions on Microwave Theory and Techniques*, Vol. 60, No. 3, 2012, pp. 845–860.

[5] Li, C., et al., "High-Sensitivity Software-Configurable 5.3-GHz Radar Sensor Receiver Chip in 0.13um CMOS for Noncontact Vital Sign Detection," *IEEE Transactions on Microwave Theory and Techniques*, Vol. 58, No. 5, 2010, pp. 1410–1419.

[6] Lee, J., et al., "A Fully Integrated 77 GHz FMCW Radar Transciever in 65 nm CMOS Technology," *IEEE Journal of Solid-State Circuits*, Vol. 45, No. 12, 2010, pp. 2746–2756.

[7] Mitomo, T. et al., "A 77 GHz 90 nm CMOS Transceiver for FMCW Radar Applications," *IEEE Journal of Solid-State Circuits*, Vol. 45, No. 4, 2010, pp. 928–937.

[8] Liu, D., and Y. P. Zhang, "Integration of Array Antennas in Chip Package for 60-GHz Radios," *Proceedings of the IEEE*, Vol. 100, No. 7, 2012, pp. 2364–2371.

[9] Person, C., "Antennas on Silicon for Millimeter-Wave Applications—Status and Trends," *Proc. 2010 IEEE Bipolar/BiCMOS Circuits and Technology Meeting (BCTM)*, Austin, TX, Oct. 4–6, 2010, pp. 180–183.

[10] Rusch, C., et al., "W-Band Vivaldi Antenna in LTCC for CW Radar Near Field Distance Measurements," *Proc. 5th IEEE European Conference on Antennas and Propagation (EuCAP)*, Rome, Italy, Apr. 11–15, 2011, pp. 2124–2128.

[11] Fonte, A., et al., "Feasibility Study and On-Chip Antenna for Fully Integrated μRFID Tag at 60 GHz in 65 nm CMOS SOI," *Proc. 2011 IEEE International Conference on RFID Technologies and Applications (RFID-TA)*, Sitges, Spain, Sept. 15–16, 2011, pp. 457–462.

[12] Gianesello, F., et al., "SOI CMOS Technology for Wireless Applications: Current Trends and Perspectives," *Proc. 2010 IEEE International SOI Conference*, San Diego, CA, Oct. 11–14, 2010, pp. 1–2.

[13] Saponara, S., and B. Neri, "Integrated 60 GHz Antenna, LNA and Fast ADC Architecture for Embedded Systems with Wireless Gbit Connectivity," *Journal of Circuits, Systems and Computers*, Vol. 21, No. 5, 2012, pp. 1–24.

[14] Simonen, P., et al., "Comparison of Bulk and SOI CMOS Technologies in a DSP Processor Circuit Implementation," *Proc. 13th IEEE International Conference on Microelectronics (ICM),* Rabat, Morocco, Oct. 29–31, 2001, pp. 107–110.

[15] Menzel, W., and A. Moebius, "Antenna Concepts for Millimeter-Wave Automotive Radar Sensors, *Proceedings of the IEEE*, Vol. 100, No. 7, 2012, pp. 2372–2379.

[16] Wang, H., "Current Status and Future Trends for Si and Compound MMICs in Millimeter Wave Regime and Related Issues for System on Chip (SoC) and System in Package (SiP)," *Proc. 2010 IEEE Topical Meeting on Silicon Monolithic Integrated Circuits in RF Systems (SIRF),* New Orleans, LA, Jan. 11–13, 2010, pp. 16–17.

[17] Wang, H. et al., "MMICs in the Millimiter-Wave Regime," *IEEE Microwave*, Vol. 10, No. 1, 2009, pp. 99–117.

[18] Scavennec, A., M. Sokolich, and Y. Baeyens, "Semiconductor Technologies for Higher Frequencies," *IEEE Microwave*, Vol. 10, No. 2, 2009, pp. 77–87.

[19] Giammello, V., E. Ragonese, and G. Palmisano, "A Transformer-Coupling Current-Reuse SiGe HBT Power Amplifier for 77-GHz Automotive Radar," *IEEE Transactions on Microwave Theory and Techniques,* Vol. 60, No. 6, 2012, pp. 1676–1683.

[20] Bin, L., et al., "Analog-to-Digital Converters," *IEEE Signal Processing*, Vol. 22, No. 6, 2005, pp. 69–77.

[21] Greshishchev, Y. M., et al., "A 40GS/s 6b ADC in 65nm CMOS," *Proc. 2010 IEEE International Solid State Circuits Conference (ISSCC)*, San Francisco, CA, Feb. 7–11, 2010, pp. 390–391.

[22] Harpe, P., et al., "A 7-to-10b 0-to-4MS/s Flexible SAR ADC with 6.5-to-16fJ/conversion-step," *Proc. 2012 IEEE International Solid State Circuits Conference (ISSCC)*, San Francisco, CA, Feb. 19–23, 2012, pp. 472–474.

[23] Mishali, M, et al., "Sub-Nyquist Sampling," *IEEE Signal Processing Magazine*, Vol. 28, No. 6, 2011, pp. 98–124.

[24] Yu, B., et al, "A 14-bit 200-MS/s Time-Interleaved ADC with Sample-Time Error Detection and Cancelation," *Proc. 2011 IEEE Asian Solid State Circuits Conference (A-SSCC)*, Jeju, South Korea, Nov. 14–16, 2011, pp. 349–352.

3

Hardware-Software Implementing Platforms for Radar Digital Signal Processing

3.1 Implementing Platforms and Performance Metrics for Radar Signal Processing

3.1.1 Implementing Platforms for Radar Digital Signal Processing

As discussed in Chapters 1 and 2, ubiquitous integrated radar applications for civil applications such as automotive driver assistance or vital signs detection need low power consumption [1–4]. Reducing the power consumption allows for the reduction of the voltage supply, of the battery size in portable devices, and of the thermal issues, thus saving cost, size, and weight of the cooling and power supply subsystems.

Beside the requirement for increased portability, the previously cited applications typically require high computational capabilities, a large data transfer rate, and large memory storage size to manage high-performance radar algorithms and techniques such those presented in Chapters 4 and 5.

As a matter of fact, radar R&D activities are characterized by the increase of signal processing tasks in the digital domain.

Recent research results on ADCs allow for power-efficient ADCs with high-end performance in terms of effective numbers of bits and sampling frequency. Therefore, the radar signal can be digitized directly at intermediate frequency, minimizing the daunting tasks to be carried out in the analog domain at radio frequency.

FFT and its inverse IFFT, time domain and frequency domain filtering [5], PC, digital down- and up-conversion, beam-forming, waveform generation by DDS, CFAR, tracking, space time adaptive processing (STAP) [6], and DOA estimation [7, 8] are some examples of common radar signal and data processing algorithms [9] to be realized in the digital domain.

Beside processing functions, the radar digital domain should be in charge of control and interface tasks, since the radar is often a node of a more complex networked system.

For example, in the automotive domain the radar subsystem should be connected to the main control vehicle network through a controller area network (CAN) or FlexRay bus interface.

Summarizing, while in the origin radar was mainly an analog system, now it is becoming a mixed-signal system [10], following the scheme reported in Figure 3.1.

This evolution from analog to mixed-signal domain is similar to what happened to the radio that now is entering the era of "software-defined radio" and "cognitive radio."

This is why some radar research projects are focusing on future digital defined architectures where the ADC moves close to the antenna and the digitization is done at higher frequencies (see Figure 3.2). The dream is that all signal processing, apart antenna impedance matching and first low-noise amplification, is done in the digital domain and the radar becomes a fully programmable/configurable device.

However, for the next years, and for the target applications of this book (automotive and health care applications of radar), the authors believe that the most suited transceiver architecture still needs a good mixer design in MMIC technology [11].

The limit for direct digitization of the signal at radio frequency is that high-speed and low-power ADC have low dynamic range (less than 8–10 effective number of bits), posing a limit to radar systems that need a wider bit range to face clutter problems.

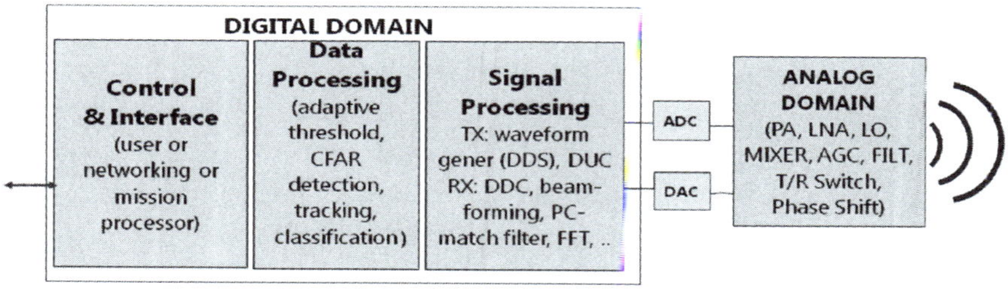

Figure 3.1 Radar as a mixed analog-digital system.

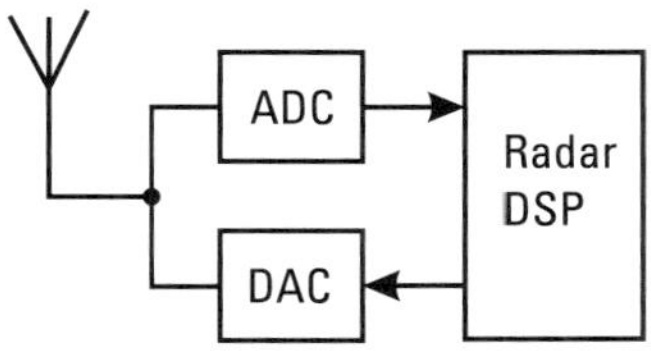

Figure 3.2 A radar with a full digital architecture will lead to an excessive power consumption for the required high sampling frequency of analog-digital and digital-analog converters and for a very high speed DSP.

Moreover, very high speed DSP at radio frequency would need very high data transfer rates and storage (large memory size), as well as high clock frequency, thus increasing power consumption.

Beside the requirements of low power and high computational/memory performance for the radar digital signal processing platform, there are market requirements such as reduced time to market, product reconfigurability, and sharing of research and design costs among several products.

This leads to stringent requirements also in terms of flexibility for implementing the radar platform. Therefore, a trade-off has to be found between the multiple requirements.

Among the possible target platforms, an increased flexibility is typically allowed by software-oriented solutions such as general-purpose processors (GPPs), DSP processors, or microcontrollers (MCUs) although at an increased power consumption versus custom hardware solutions must be taken into account.

Minimization of power consumption is the main feature of hardware-oriented solutions such as application-specific integrated circuits (ASICs) [12–14] and FPGA.

The scenario of the possible radar implementing platforms is even more complex since new software-oriented or embedded hardware-software platforms are emerging on the market such as:

- The GPU, which is used more and more as high-performance massively parallel computing units for DSP and not only for imaging applications.
- The field programmable system-on-chip (FPSoC), merging the hardware capability of FPGA with the software flexibility of soft or hard programmable cores integrated in the same chip.
- The multiprocessor system on chip (MPSoC), where multiple low-power programmable cores can ensure the required computational capability for real-time radar processing although at a better power efficiency than GPPs.

It is worth noting that typically external analog to digital converters and digital to analog converters are required for the intermediate-frequency parts of a radar transceiver, since GPP, GPU, DSP, and FPGA platforms are equipped only with digital interfaces.

The only solutions that generally integrate the ADC, and in some case also the DAC, are custom IC designs or microcontrollers.

3.1.2 Main Performance Metrics for Radar Implementing Platforms

There are several performance metrics to be considered to compare different possible implementing platforms. A fully detailed analysis will require the consideration of hundreds of parameters.

To enable a fast comparison of different competing platforms, we suggest the extraction of few "key" parameters to characterize each candidate solution for radar implementation. In most cases a trade-off is to be found since a single platform optimizing all parameters does not exist. However, when the number of different parameters to be considered is reduced to few "key" values, the comparison can be easily realized adopting classic multiobjective analysis tools such as the Pareto plots or the spidermaps.

Hereafter, we briefly review the list of suggested key parameters to be considered when analyzing an implementing platform for radar digital signal processing.

3.1.2.1 Computational Capability

In the literature, computational capability is typically measured in terms of giga operations per second (GOPS). However, it is not easy comparing platforms with different instruction set and architectures, just in terms of GOPS: as an example, reduced instruction set computer (RISC) computing platforms typically need multiple instructions to execute a complex operation that a complex instruction set computer (CISC) processor can execute in few cycles, or even in just one machine cycle.

Hence, using only the GOPS figure of merit to compare different radar implementing platforms has the risk of overestimating the performance of a RISC computing core versus that of a more complex CISC one. To avoid this risk, in the literature the computing platforms are typically characterized with reference to common benchmarks.

The most used benchmark is the Dhrystone. It is a synthetic computing benchmark program, first developed in 1984 by R. P. Weicker intended to be representative of system (integer) programming. The Dhrystone grew to become representative of general processor performance, and currently the Dhrystone v 2.1 is the most used one. This is why most of processors are today characterized in terms of DMIPS (Dhrystone millions of instructions per second) or DMIPS/MHz.

The DMIPS figure for a given computing core represents its relative speed versus a VAX 11/780 (a particular "1 MIPS" machine) computing core selected as reference. For example, if a given processor completes the Dhrystone benchmark 200 times faster than the VAX 11/780 does, then it would be considered a 200 DMIPS machine. For processors that can be run at various clock frequencies, often the value is divided by MHz (e.g., if the 200 DMPIS processor runs at 100 MHz, the value is 2 DMIPS/MHz).

For example, the StrongARM SA-110 processor at 214 MHz has a 240 DMIPS capability (i.e., roughly 1 DMIPS/MHz). The StrongArm family was firstly developed from DEC, then sold to Intel, which replaced it with the xScale family of processors that recently passed to Marvell.

The DMIPS scores of the PJ1 and PJ4 processors from Marvell are 1.46 DMIPS/MHz and 2.41 DMIPS/MHz, respectively.

The famous ARM Cortex A9 core has a capability of 2.5 DMPIS/MHz (in the market this core is implemented in many products of the main semiconductor industry leaders with clock frequencies in the order of 1–2 GHz) whereas the Cortex-A8 achieves 2.0 DMIPS/MHz and the Qualcomm's Snapdragon reaches 2.1 DMIPS/MHz.

On the contrary, microcontrollers have much lower absolute performance. For example, those of the Micron PIC family are limited to few tens of DMIPS: 10 DMPIS for a PIUC18F8, 40 DMIPS for a dsPIC33FJ.

Although largely used in the literature, the authors of this book believe that the DMIPS figure is not enough for a fair comparison of different computing platforms for radar applications. Indeed DMIPS is a good benchmark for general purpose computing applications, mainly based on integer arithmetic, while radar computing involves application-specific operations with floating-point operands and with a frequent use of specific arithmetic functions such as matrix calculation, Fourier transform, FIR/IIR filtering, and so on.

This is why we suggest considering, beside the DMIPS or DMIPS/MHz figure, also specific benchmarks in terms of DSP-dedicated instructions, at least the giga multiply and accumulate operations per seconds (GMACs) or the time needed to execute a well-known DSP algorithms like the fast Fourier transform.

3.2.1.2 Operand Bit-Width and Arithmetic Type

Beside the number of operations executed per second, when comparing different radar computing platforms, the supported operand bit-width and the arithmetic type must be considered, too.

The authors believe that for a good processing core, at least 32-bit width is mandatory, and that having a floating-point coprocessor is useful to face the large dynamic range of radar signals including clutter.

3.2.1.3 Memory

Radar signal processing is a typical data-dominated application involving a large amount of data transfers between the processing core and the memory subsystem with large operand size. Some emerging radar techniques such as space time adaptive processing are particularly heavy in terms of required load/store operations and size of the operands requiring high memory access frequency and high memory storage size.

High access speed and high density storage are often conflicting requirements; therefore, the proper design of a memory hierarchy between the computing core and the large off-chip background data memories is mandatory for the success of radar applications.

Fast static RAM memories operating on-chip close to the processor have lower storage density than dynamic RAMs operating off-chip as modules mounted on board or integrated on-chip in modern system-on-chip technologies, or than hard disks (realized with solid state or still with classic magnetic technologies).

Therefore, multiple levels of on-chip SRAM-based caches are envisaged to exploit the capabilities of the DSP computing core. The storage capabilities of on-chip SRAM caches are usually limited on the order of megabytes. For larger storage requirements, up to gigabytes, further levels of synchronous DRAM modules are needed. Recent advances in semiconductor technology allow the integration of embedded DRAM modules in the same chip with the processing core and the SRAM, so that multiple layers can be envisaged also for the DRAM (e.g., a first DRAM layer integrated on chip, faster, and a second DRAM layer of chip, characterized by a large storage density). Terabytes storage applications need off-chip hard disk capabilities that can be realized today with solid-state technologies (EEPROM memories) instead of using magnetic hard disks. The use of EEPROM devices for terabyte storage minimizes access delay, size and weight of the memory subsystems.

The new frontier in memory technology is represented by the research of memories allowing for nonvolatile capabilities like PROM but fast access like in RAM.

3.2.1.4 I/O Interfaces

Since radar transceivers are parts of more complex systems (e.g., driver assistance and collision avoidance in automotive applications), usually involving networking with other units, the richness in terms of input-output (I/O) interfaces is another important aspect when evaluating and selecting radar implementing platforms.

When selecting a platform the designer should consider the data rate, data type, and protocol supported by the I/O interfaces and the user interface available on the host unit that remotely controls the sensing node.

Typically a low rate serial interface (RS-232, I2C, or SPI) is useful as a configuration link between the radar unit and a host central processor. For transfers of larger data size, higher throughputs are needed: the trend for civil applications is exploiting USB or Ethernet connections, although in automotive applications CAN, with data rates up to 1 Mbps below 40m, is the most used backbone connection protocol.

In measurement applications where the radar sensing unit can be seen as a node connected to a host control, the serial link RS-485 can be used (as an extension of RS-232, offering data transmission speeds of tens of Mbps at 10m, with a rule of thumb that the speed in bits per second multiplied by the length in meters should be in the order of 10^8) with a Labview GUI at host side.

3.2.1.5 Power/Area Efficiency

For military applications, power consumption and size of the computing platform are not key issues, but for civil applications, the power and area costs of the radar computing kernel are main performance metrics.

For example, in automotive applications the radar driver assistance platform has to be mounted on board, without reducing space for passengers or reducing aerodynamic car performance, and it must be supplied by the car electrical system with minimal overhead.

This is why, beside the already cited MIPS/MHz cost figure, the MIPS/mW and MIPS/mm^2 cost metrics should be considered when selecting the target implementing platform (or, similarly, GOPS/MHz, GOPS/mW, and GOPS/mm^2 values).

3.2.1.6 Flexibility

Unlike the other metrics, it is difficult to summarize the flexibility level of a platform with a single numerical value. With the term flexibility we refer to the capability of the program of a platform to change (software reprogrammability) and/or change the configuration of the hardware resources of the platforms, (e.g., hardware reconfigurability typical in FPGA devices of FPSOC platforms).

Hardware reconfigurability involves the reconfiguration of logic blocks or connections between logic blocks and memories and I/O interfaces or connections between several logic blocks to enable new and different capabilities in the platform. As we will further discuss in the rest of the chapter, since flexibility (programmability or reconfigurability) comes with an increased cost in terms of power or area overhead, and considering that radar systems involve application-specific tasks, the authors believe that the most suited platforms for radar computing should be searched between those allowing for programmability or reconfigurability in a class of algorithms (GPU, DSP, FPGA) and the non-general-purpose ones (GPPs, microcontrollers). Similarly, those approaches with poor flexibility (ASIC) are not suitable.

3.2 Hardware-Software Architecture for a Cost-Effective Radar

The approach when designing a custom IC for radar signal processing can provide the best in terms of performance for a given technology at minimum area occupation and power consumption. Moreover, digital and mixed analog-digital (such as analog-digital and digital-analog converters) functions can be integrated together in the same chip using CMOS technology. To increase the flexibility of the ASIC approach in the same chip, programmable cores can be integrated realizing a complete SoC.

However, design and validation time and costs of a custom IC are high. Since the ASIC design approach is characterized by high nonrecurrent costs, the cost per produced chip can be minimized only for large volume markets. Today, the radar system concept for bio, automotive, or other ubiquitous applications is still changing from one platform generation to the next. Automotive radar is not yet a commodity application with accepted and frozen standards, and hence the market is still not a large-volume one.

Considering that the number of radar units for specific civil applications is still limited, if compared to other large-volume markets as smart phones, tablets, or networking devices, in next years an architecture based on MCU plus DSP or FPGA is considered more suited than designing custom IC for radar. The MCU core will act as low-complex controller while the DSP or FPGA will act as a coprocessor for computing intensive radar signal processing applications.

Thanks to the availability of a library of reusable intellectual property (IP) cores, a radar platform design, based on MCU and FPGA mounted on a single-board, can be seen as a prototyping step toward a future migration to SoC if and when radar becomes a commodity and its market becomes a large-volume one.

It should be noted that IP cores are predesigned macrocells for processors, memories, custom DSP functions, and I/O interfaces that can be configured and synthesized in a given silicon technology. As an example, most vendors of FPGA and FPSoC devices such as Xilinx, Altera, Microsemi, Cypress, and others offer (directly or through partner companies specialized in the design of IP reusable cores) a rich library of reusable macrocells for most functions required in a radar platform developments: controller cores, specialized DSP functions for FFT evaluation, digital signal filtering, digital beam-forming, and so on.

With reference to automotive radar applications, Figure 3.3 presents a possible scheme for a radar computing platform realized as a single board system. The figure highlights the following key units, realized as discrete components mounted on the same board: an antenna array with radio frequency transceiver (power amplifier at transmitter side and multichannel receiver chain) realized with MMIC technology [15]; power supply unit; ADC operating at intermediate-frequency; and radar digital signal processing tasks implemented

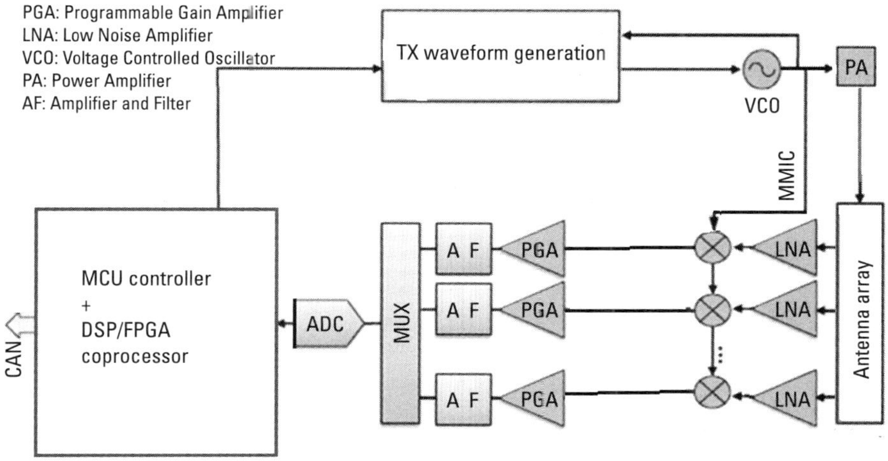

Figure 3.3 Automotive radar main blocks.

on FPGA while the main controller functions and interfacing, based on CAN, with the main vehicle network are implemented on a MCU. The antenna is conceived as an antenna array since for estimating the direction of arrival at least two antennas for each direction are needed; four antennas are needed for the estimation in both azimuth and elevation planes.

The MCUs available in the market, differently from GPPs, are low-cost devices with reduced power consumption. MCUs are typically characterized by low computational capabilities, in the order of MIPS rather than GOPS.

Moreover, they support integer operations only and are equipped with low data rate digital I/O. As far as memory hierarchy is concerned, MCUs integrate on-chip different types of memories (both volatile and nonvolatile storage units such as SRAM and EEPROM), but the total amount of on-chip memory is small (few to tens of kilobytes).

Being limited both in data and instruction memory, a MCU has limited capability of handling complex programs and large set of data.

The MCUs for their reduced cost, size, and power consumption, and their software programmability, are the core of ubiquitous electronic systems (automotive ECUs); however, for radar applications MCUs are suitable only for tasks such as control of data/operation flow but not for DSP operations.

This is why in the envisaged automotive radar platform (see Figure 3.3), the MCU acts as a baseband control unit and manages the telemetry/telecontrol on vehicle network (CAN or FlexRay bus), but a DSP or FPGA co-processor is required for computing intensive signal processing operators like pulse compression, filtering, FFT, and CFAR.

An example of a commercial realization of a radar platform following the concepts we envisage in Figure 3.3 is the third generation of long-range automotive radar recently realized by BOSCH (called LRR3) [16] and reported in Figure 3.4.

In the solution proposed by BOSCH for long-range automotive applications, the whole radar platform is implemented as a main board, hosting the power supply unit, the interface toward the CAN vehicle networks, and the processing unit (controller tasks and digital signal processing tasks). In the LRR3 solution the radio-frequency and the intermediate-frequency interfaces and the printed antenna array (four antennas) are realized with monolithic millimeter wave components mounted on a dedicated daughter board.

The next paragraph will discuss why the authors do not consider general-purpose processors a suitable implementing platform for radar in ubiquitous civil applications and the alternative solutions for digital signal processing using DSP (SIMD architectures or massively parallel computing platforms such as GPU) or FPGA, high-end devices, or cost-effective ones.

3.3 DSP and GPU for Radar Signal Processing

In the last years, GPPs are evolving as software-programmable computing architecture with multiple cores, large operand size (64 bit or higher), and tens of GOPS of computational capabilities.

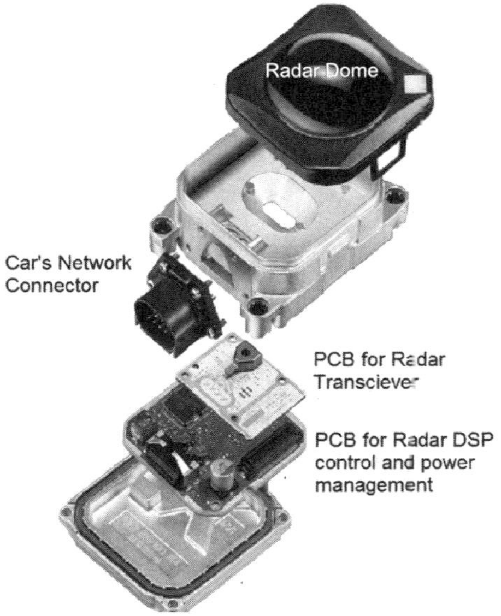

Figure 3.4 FMCW long-range automotive radar from BOSCH—third generation.

The memory subsystem of modern GPPs is characterized by multiple cache levels with at least several megabytes of on-chip storage capabilities: in multicore systems at least two levels are dedicated to each core (L1 and L2 caches), while the L3 cache is shared among the multiple cores.

Fast connections toward off-chip DDR dynamic RAM memories and fast I/O interfaces are also present in modern GPPs. Notwithstanding that GPPs can substitute MCU + FPGA, managing both control tasks and computing intensive signal processing operators, the energy efficiency of general-purpose processors (usually << 1 GFLOPS/Watt) and the absolute level of power consumption, in the order of tens of Watts, are far from the specification of highly integrated low-power radar applications.

Better energy efficiency, but still achieving real-time processing of radar computing-intensive tasks, can be achieved with digital signal processors. They are software-programmable processors with instruction set optimized for signal processing rather than general-purpose computing.

As an example, the arithmetic logic unit (ALU) of a digital signal processor is typically equipped with hardware resources for fast multiply and accumulate (MAC) implementations and automatic address generation (AGU), with both fixed or floating-point versions available. Data sizes are at least 32 bits, and different architectural approaches are available, with the most diffused being single instruction multiple data (SIMD), executing the same instruction in parallel on multiple data, or very long instruction word (VLIW) capable of executing in parallel multiple instructions on multiple data.

Digital signal processors can be single core or multiple cores platforms; an evolution of this class of processors as massively parallel computing platforms is the GPU.

Digital signal processors allow for high-end computing performance, up to tera floating-point operations per second in case of GPU, a value much higher than multicore GPP, and with a better energy efficiency than GPP in case of the execution of signal processing tasks.

Digital signal processors for consumer, automotive, and biomedical markets are already available from different vendors such as Texas Instruments and Analog Devices.

The software development environment of a digital signal processor typically allows the design of programs with high-level languages such as C and provides a rich set of libraries and all the CAD tools needed for debugging, compiling, linking, assembling, and the final programming.

Some digital signal processor architectures are equipped also with embedded operating systems, often with real-time capabilities such as a real-time operating system (RTOS) (e g., the DSP-BIOS RTOS for the DaVinci family of Texas Instruments, µC/OS-II™ Real-Time Kernel for the Blackfin processor family of AnalogDevices). Lots of digital signal processors today integrate an

ARM core and support embedded linux or windows CE or android operating systems.

Digital signal processors are often available as reusable IP cores to be integrated in custom SoC: soft macrocells for which a parametric HDL description exists, which can be modified and configured by the core user and then synthesized in the target technology, or hard macrocells, already synthesized in the target technology by the IP vendor, and not customizable by the IP user; that is, integrated as a black box in the back-end phase of the IC design.

As an example of possible digital signal processors to be adopted for radar signal processing, in the rest of this chapter we will analyze three different architectures:

- Multicore cell DSP from IBM using as a basic core a synergistic vector processor representative of a SIMD architecture;
- The Fermi family from NVIDIA, as an example of massively parallel GPU architectures and their use for computing intensive DSP tasks;
- DSPace core developed by an European consortium (Universities of Pisa and Aachen and industries like Intecs, Sitael, and SpaceApp), which is an example of a VLIW design, targeting avionic/space signal processing applications, but usable also for ground applications.

3.3.1 Vector DSP and the CELL Many-Core Computing Engine

The synergistic processor of CELL from IBM is a vector processor (i.e., it has a SIMD architectures operating a single instruction each time on vector of data). It is a 128-bit processor with 128 special registers each 128 bits wide. The synergistic processor can operate with floating-point or integer data with a computational capability of 25.6 GFLOPS when operated with a single precision on floating-point data. When integer arithmetic is used, the computational capability of a synergistic vector DSP increases to 51 GOPS. This processor has also 256 KB of local SRAM memory and has direct memory access (DMA) instructions for moving data between local store memory and main memory.

The synergistic processor is integrated as a signal processing core in the multiprocessor of IBM called cell broadband engine (BE) [17, 18]. The CELL BE integrates one 64-bit power processor element (PPE) core in charge of operation scheduling and data flow control, and eight synergistic processors (SPs) dedicated to DSP computing and working according to a SIMD scheme. The nine processors are interconnected through a common on-chip communication link with ring topology. The on-chip network has a communication capability of roughly 200 GBps sustained bandwidth. The CELL BE has also dedicated resources for off-chip DRAM memory interfacing and for fast I/O off-chip communication; the relative driving unit is also connected to the ring high-bandwidth network.

The Cell BE has 256 KB of local memory for each SP, 32 KB of L1 data and instruction caches for the PPE, and 512 KB L2 shared cache.

Each SP core has a transistor count of 21 millions, an area of 14.77 mm^2, and a power cost of 5W in 90-nm CMOS technology. Each SP has an area of 11.08 mm^2 and 6.47 mm^2 in 65-nm and 45-nm technology, respectively. The multicore CELL BE has an area of 235 mm^2 in 90-nm CMOS technology, 174 mm^2 in 65-nm CMOS, and 115 mm^2 in 45-nm CMOS SOI [19, 20]. Therefore, in submicron CMOS technology, by scaling the same architecture from 90-nm to 65-nm and then to 45-nm technology node, the overall processor area scales with a factor roughly $\lambda 2/\lambda 1$ with $\lambda 2$ and $\lambda 1$ being the channel lengths of the MOS devices in the two technologies. In the past, with larger technology nodes above 100 nm, such a ratio was $(\lambda 2/\lambda 1)^\theta$ with θ a fitting parameter between 1 and 2.

The power consumption of the CELL BE at 3.2 GHz is up to 80W in 90-nm CMOS technology, while at the same clock frequency in 45-nm CMOS technology is up to 40W [19, 20].

The overall computational capability of the CELL BE computing platform at 3.2 GHz is 205 GFLOPS (floating-point arithmetic) or 410 GOPS (fixed-point arithmetic) when realized in 90-nm CMOS silicon-on-insulator (SOI) technology. Therefore, the computational capability per clock frequency is on the order of 128 MIPS/MHz in fixed-mode and half in the floating-point mode. With respect to the similar performance figures discussed earlier for CortexA9, the CELL BE is at least 50 times faster and is 100 times faster than the StrongARM. This performance enhancement is the result of the combined use of multiple processors (9: 8 vector DSP + 1 power PC core) and the fact that each of the eight synergistic processors is optimized for SIMD signal processing functions.

In 65-nm and in 45-nm CMOS SOI, the clock frequency of the CELL BE can rise up to 6 GHz and the peak computational throughput is roughly 380 GFLOPS.

The computational capability enhancement of the CELL is paid in terms of area and power consumption overhead versus less powerful ARM cores; tens of watts for the CELL BE versus few watts or even less than 1W for ARM-based computing platforms.

At the University of Pisa, we benchmarked the CELL BE at 3.2 GHz versus different general purpose processors, considering a function that is often used in signal processing applications, and particularly in radar imaging, such as fast Fourier transform (FFT). For the FFT benchmarking, we considered different length size from 1 K to 64 K.

The target reference processors for the comparison with the CELL BE are an INTEL Pentium at 3.6 GHz with 2 MB of L2 cache, an AMD Opteron at 2.4 GHz, and a Freescale 744 MX processor at roughly 1 GHz.

The results of the benchmark are reported in Table 3.1, with a relative performance of 1 being that of the Freescale 744Mx processor. The results of the table clearly show that CELL BE is at least one order of magnitude faster when implementing digital signal processing functionalities.

3.3.2 GPU

GPUs are now evolving as massively parallel platforms for graphics but also for computing-intensive signal processing algorithms with a high degree of parallelism [21–23]. For example, the Fermi architecture of NVIDIA is composed of 512 compute unified device architecture (CUDA) processing cores hierarchically organized in 16 streaming multiprocessors, each with 64-KB L1 cache and 128-KB local register file, and sharing a common 768-KB L2 cache. The total amount of on-chip memory is roughly 4 MB.

The Fermi GPU architecture also has a PCIe host interface and six 64-bit double data rate (DDR) DRAM interfaces. Each CUDA core is capable of both integer and floating-point operations with 64-bit results.

Each of the 16 streaming processors also has four 40-bit special function units for fast approximation of nonlinear operators (square root, sin, cos, exp, log functions) and 16 load/store units. The Fermi NVIDIA GPU leads to high computational power, up to 1500 GFLOPS in single-precision, orders of magnitude higher than general purpose processors such as Core 2 or Atom.

Table 3.2 reports a complete comparison between some GPU models (ATI Radeon or NVIDIA Fermi) and some GPPs platforms in terms of computational capabilities (measured in GFLOPS); power consumption (measured in watts), sustained off-chip communication bandwidth (GBps) and some efficiency metrics as GFLOPS/watt.

Although real-time processing is not an issue for such GPUs, their area and power consumption is suited only for desktop applications and workstations, not for handheld or mobile devices. Hence GPU-based radar signal processing

Table 3.1
Speed-Up Factor of the CELL BE Versus Reference Processors When Executing FFT with Different Lengths

	1 K-point FFT	8 K-point FFT	64 K point FFT
CELL BE	32	56	30
Pentium	1.5	2.1	1.9
Opteron	0.9	1.3	1
Freescale	1	1	1

Table 3.2
Comparison of Some GPU Models Based on NVIDIA Fermi or ATI Radeon Architecture and Some GPPs from INTEL

GPP	GFLOPS	Bandwidth (GB/s)	Watt	GFLOPS/ Watt	GFLOPS/ Bandwidth
Intel Core2Duo E8600	27	10.4	65	0.4	2.6
Intel Core2Duo SP9600	20	8.33	25	0.8	2.4
Intel Core i7-870	47	22.89	95	0.5	2.1
Intel Core i7-820QM	28	21	45	0.6	1.3
GeForce GT 240M	174	25.6	23	7.6	6.7
GeForce GTS 250M	360	51.2	28	12.9	7.0
GeForce GTX 260	875	112	171	5.1	7.8
AMD/ATI 5870	2720	153.6	188	14.5	17.7
AMD 5750	1008	73.6	86	11.8	13.7
AMD 5770	520	51.2	30	17.3	10.2
AMD 5830	800	25.6	24	33.3	31.3
AMD 5870	1120	64	50	22.4	17.5

can be used for airborne or naval applications, or for surveillance applications from a remote control room but not for on-board automotive applications or handheld vital sign detection devices.

For example, the GTX480 Fermi NVIDIA, realized in 45-nm CMOS technology, has a transistor count of 3 billions, an area occupation for the core of 422 mm^2, and power consumption higher than 150W when implementing algorithms with a computational workload higher than 500 GFLOPs. This power consumption requires the use of active cooling systems to keep under control the operating temperature of the GPU chip.

Scaled versions of CUDA-based GPU exist (e.g., the NVIDIA Quadro 4000 GPU has 256 CUDA cores, four 64-bit DDR DRAM interfaces and a maximum power consumption of 100W).

The massively parallel approach has been followed also by other designs available in literature (e.g., the TILERA architecture). The Tile 64 processor realizes in CMOS 90 nm an array of 64 basic processors with a 64-bit instruction set and 32-bit integer ALU communicating through a 2D mesh on-chip interconnect. The Tile 64 reaches a computational throughput of 144 integer GOPS (32-bit operations) and 384 integer GOPS (8-bit operations) with a 750-MHz clock. Its complexity amounts to 615 millions of transistors; the area is 433 mm^2, 215 mm^2 in 45 nm, and the power consumption at full load is roughly 11W.

Although a comparison of hardware/software platforms with different architectures, instruction sets, and computing performances is difficult to implement, Figure 3.5 provides a visual representation of the trade-off between die

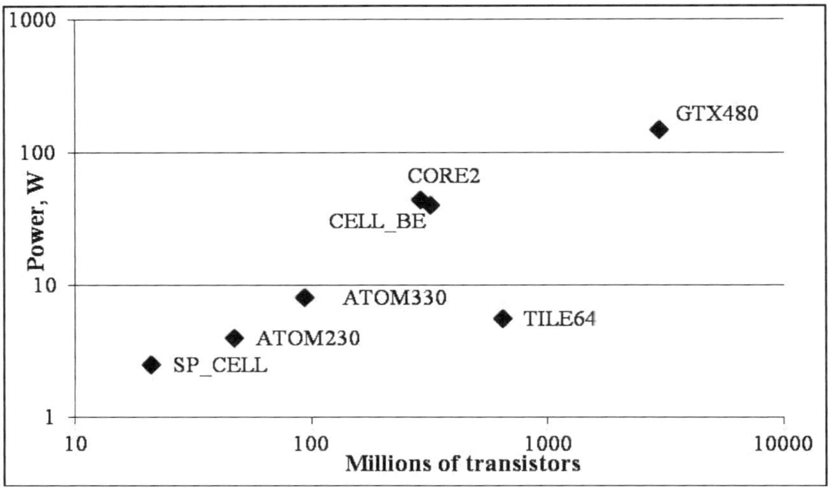

(a)

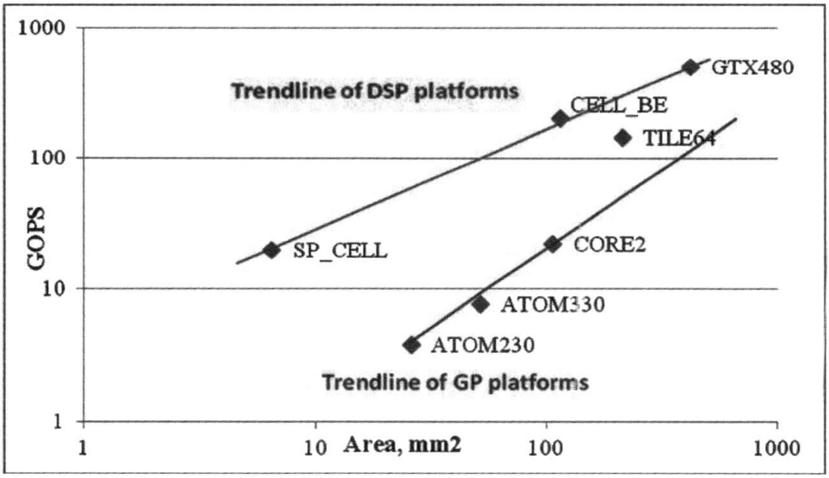

(b)

Figure 3.5 Area, number of transistors, computational power (GOPS), and power consumption of state-of-the-art single-core, ATOM 230, and 1 synergistic processor (SP) of the CELL BE and multicore systems, ATOM330, Fermi GPU GTX480, Core2, CELL BE, and Tile64.

size (due to circuit complexity spent for increased parallelism and hence increased performance), power consumption, and computational capabilities in state-of-the-art single-core and multicore systems.

Reported data refer to platforms realized in 45-nm silicon technology nodes and to operations on 32-bit data (integer or single-precision floating-point).

Figure 3.5 highlights that when considering as performance metric the ratio GOPS/mm^2, there are two different trend-lines: one for platforms optimized for general purpose computing applications (e.g., Atom, Core2) and one for platforms optimized for DSP applications. When the level of parallelism increases (e.g., from 1 in the SP_CELL to many cores in the CELL-BE or Tile64 or GTX480), the difference between the two trend-lines is reduced since at high parallelization levels the performance bottleneck is represented by communication and the computational power increases slowly versus the number of cores.

At the University of Pisa, we implemented a benchmarking of a GPU computing platform versus a general purpose processing platform considering some key algorithms for radar signal processing. The considered GPU processor is a NVIDIA GTX260, which has been tested against a Core2Duo GPP E8600. The considered radar DSP algorithms are as follows:

- FFT and IFFT;
- Timed domain FIR (TDFIR) and frequency domain FIR (FDFIR) filters, the latter realized according the following chain: FFT cascaded to a TDFIR cascaded to a IFFT;
- Single value decomposition and QR-decomposition on a complex large size matrix;
- CFAR.

Similar benchmarking activities have been done in the literature by other academic and industrial R&D group such as SAAB [24]. The results achieved by SAAB and reported in the literature are similar to those reached in Pisa and reported hereafter. SAAB also considers in its benchmarking analysis STAP techniques. STAP has been not considered in our benchmarking activity since this technique is too computationally complex for the target low-power ubiquitous civil radar system.

Considering the different types of radar algorithms, the GPU resulted in a speed-up factor from 10 percent up to two orders of magnitude versus GPP, depending mainly on the size of the input array data, on the regularity and parallelization degree of the data flow, and on the bottleneck caused by memory transfers. Indeed in most cases the whole GPU computing platform including the memory system has much worse performance than those achievable by the GPU kernel alone.

From our benchmarks for linear filtering in time domain, the GPU kernel has a speed-up up to 70 times versus the considered GPP unit; the higher the length of the input array data, the higher the GPU speed-up.

Considering the whole computing platform, including the memory transfers between the computing kernel and the large off-chip memories, the speed-up factor is reduced.

Considering a CFAR algorithm, the GPU kernel has a speed-up versus the GPP platform from a factor 3 times to a factor 30 times, depending on the data set used for the benchmark. When also considering memory transfers, the speed-up factor is halved.

Finally, considering FFT (the results for IFFT are similar), the speed-up factor of the GPU versus the GPP is up to 40 times. The higher the number of the input vector length, the higher the gain. When also considering the memory subsystem, the gain factor is reduced by a factor 4 times.

3.3.3 VLIW DSP for Space Applications (DSPace) Processor

In the context of the European project DSPace, a VLIW architecture has been recently released targeting low power consumption and a signal processing capability of around 1 GFLOPS peak. The core is released with intellectual property macrocells that can be fitted in FPGA or integrated in SoC deigns, together with a correlated a high-quality software development environment (SDE).

With respect to the previously analyzed DSP platforms (vector DSP and many core CELL architectures), the DSPace design is characterized by a lower computational capability and lower power consumption, being more suited for the ubiquitous civil radar applications of this book, although it cannot support complex radar algorithms like STAP. Although the main DSPace target is aerospace application, the processor can be used also for ground DSP applications.

The cardinal set of system requirements for the DSPace processor design was the new generation of space-qualified DSP identified by the European Space Agency [25]. Hereafter we list the main ESA requirements:

- Processing power ≥ 1000 mega instructions per second (MIPS);
- Radiation hardness, total ionization dose ≥ 100 krad;
- Memory protection with error detection and correction (EDAC);
- Support for aerospace standard I/O interfaces;
- High-quality software development environment;
- No access restrictions for European users (i.e., no ITAR restrictions);
- Reduced power consumption.

To achieve a significant DSP performance increase, while maintaining a moderate clock frequency and a hence a moderate power consumption, the processing core is based on a parallel architecture. Second, to maintain the functionality in radiation environment, the DSP implements by design radiation hardening techniques: triple modular redundancy (TMR) with voting logic to protect internal registers as well as EDAC protection for on-chip memories to correct single error and to detect double error events (SEC/DED).

The processor offers also general purpose capabilities, and hence it can act as main control unit, but at the same time it provides a lot of specific and optimized instructions to speed up the recurrent operations in signal and image radar processing (i.e., jump with decrement of a register, bit reversal, circular addressing mode).

Finally, the DSPace design has a complete SDE, with C compilers, optimizer, assembler, linker, and instruction-level simulator, to enable the software architects to easily program the device. All the development tools, except the C compiler, come from a concurrent hardware/software design approach. Indeed, the compiler is based on the reuse of a mature open-source GCC-based toolchain for ground DSP, ensuring binary compatibility through the translation and optimization of GCC assembly into DSPace assembly code.

As far as the DSPace architecture is concerned, a minimal and complete set of building blocks are arranged in order to supply the data processing unit (DPU) with the program instructions and the data to be elaborated, allowing also the possibility to communicate with the hosting environment. The DPU represents the computational element of the chip. Moreover, there are some on-chip peripherals able to provide I/O capabilities for operative and control purposes. The system works with cache memories both for instructions and data, aiming to improve its performance. All the communications with the external main memory (i.e., in case of cache miss) are driven by the direct memory

Table 3.3
Speed-Up Factor of GTX260 GPU Versus Core2Duo GPP for a Time Domain FIR Implementation, Computing Kernel Alone

		Length of the Input Vector					
		1K	**4K**	**16K**	**64K**	**256K**	**1M**
FIR filter length	**32**	7	18	34	48	72	77
	64	8	24	43	49	65	74
	128	10	28	44	52	61	74

Table 3.4
Speed-Up Factor of GTX260 GPU Versus Core2Duo GPP for a Time Domain FIR Implementation, Computing Kernel Plus Memory System

		Length of the Input Vector					
		1K	**4K**	**16K**	**64K**	**256K**	**1M**
Filter length	**32**	1,5	3	5	4	11	17
	64	2	4	7	12	19	26
	128	2	5	10	19	26	37

access (DMA) controller through the generic memory controller able to deal with double data rate (DDR2) RAM memories. The two SpaceWire interfaces, which are the natural choice in the space field, allow for the information exchange between the DSPace chip and the hosting system (i.e., read the status of a sensor, actuate a control, or output a result). SpaceWire can be also used for high data rate ground communications where high reliable links are required as demonstrated in [26] for the automotive, defense, or physics experiment application fields. AMBA standard busses are finally involved within the chip to connect the building elements. The high-level DSPace [27] SoC blocks diagram is shown in Figure 3.6.

The DPU is based on a VLIW architecture featuring eight computational units organized in parallel, allowing the performance of up to 8 RISC instructions at every clock cycle. Assuming an operational frequency of 125 MHz, it ensures near 1000 MIPS peak performance as expected.

Figure 3.7 shows in detail the internal structure of the DPU. This is the processing block of the DSPCE SoC, in charge of fetching, decoding, and executing the program instructions coming from the instruction cache (I-cache) elaborating the data provided by the data cache (D-cache) or by the SpaceWire ports.

The DPU data path consists of the following:

- A register file with 64 general purpose 32-bit locations;
- Four instances of arithmetical-logical unit (FP_ALU) capable of all the logic, arithmetic (with the exclusion of multiplying), conversion, and I/O operations;
- Two multiplier unit (FP_MUL) dedicated to multiplying operations;
- Two address generation unit (AGU) for accessing the memory;
- Two 64-bit ports (LD) to load data from the memory to the register file;
- Two 32-bit ports (ST) to store data contained in the register file to the memory.

All the functional units work with immediate operands or values provided from the register. Supported operands are compliant with the 32-bit 2s complement fixed-point and IEEE 745 single precision floating-point formats. Indeed the HW support for extended fixed-point (40 bit) or double-precision floating-point (64 bit) have not been considered in this version. Table 3.5 summarizes the capabilities of each kind of computational unit.

The DPU is designed with a single-cycle throughput, and it implements a seven-stage pipeline architecture. The fetch phases contains three stages of the pipeline. It is in charge of retrieve instructions from the I-cache according to the

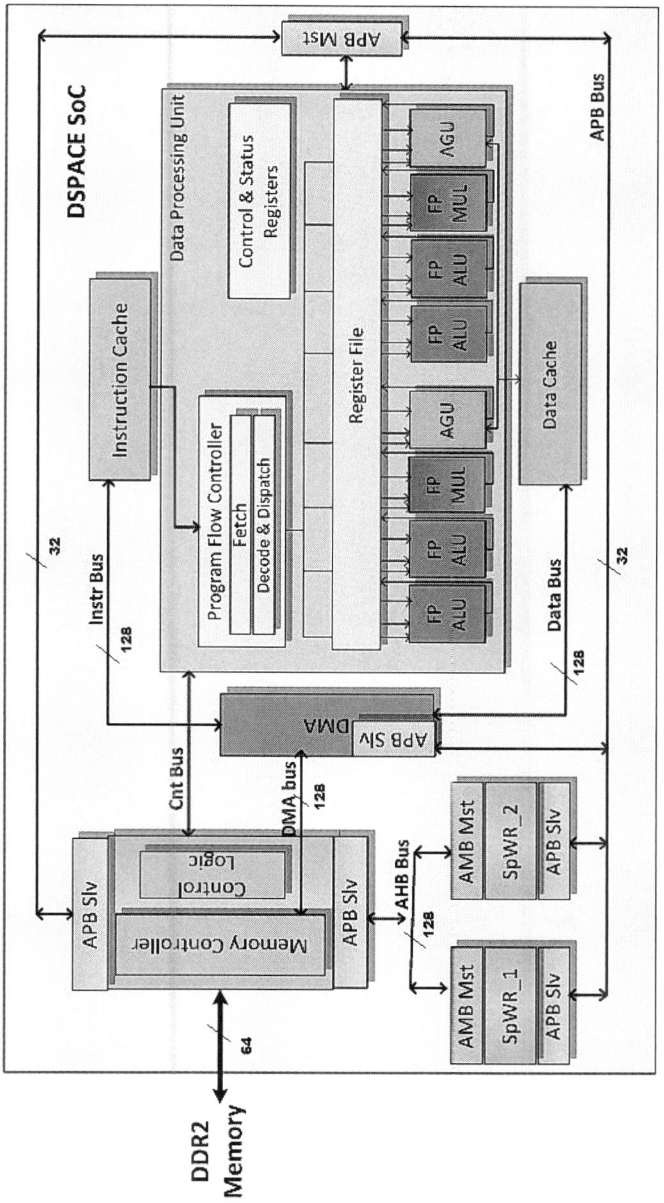

Figure 3.6 DSPace complete block diagram.

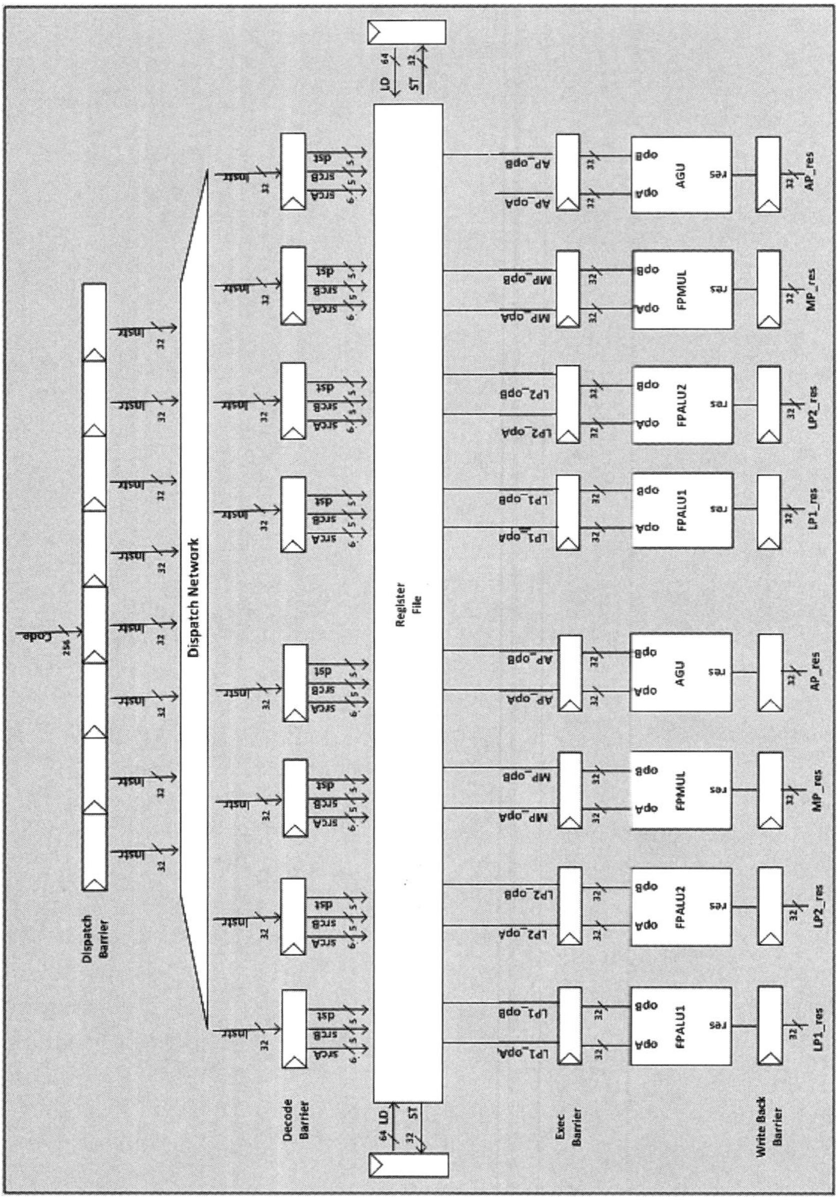

Figure 3.7 Block diagram of the internal structure of the DPU of the DSPace processor.

Table 3.5
DSPace Computational Units Capabilities

FP_ALU	FP_MUL	AGU
32-bit logical op 32-bit arithmetic op 32-bit compare op 32-bit bit-field op 32-bit shifts Min/max op Normalization Bit reversal Trailing bit count Single-precision floating-point compare op Single-precision floating-point arithmetic op Min/max floating-point op Integer/floating-point conversions Branch Branch with register decrement Interrupt handling Control register transfers to/from register file Input/output transfers	32- x 32-bit multiply op Single-precision floating-point multiplication 16- x 16-bit multiply op	32-bit addition and subtract in linear or circular address calculation Load and store with 5-bit and 15-bit constant offset using 18 different addressing op Load and store nonaligned bytes, half-words, words, and double-words

program flow. The following two stages of the pipeline (decode phase) provide the source operands decoding and the dispatching of the instructions to the right execution unit. During the execution stage, the units perform the assigned instructions and then provide the results of the computation in the register file in the final write-back stage. The pipeline is stalled in case of cache misses, both for instructions and data, until the entire block the missing data belong to has been transferred from the external main memory to the involved cache via the DMA controller.

The DPU works with instruction packets (IPs) composed of eight instructions. The instructions within each IP that can be executed in parallel in the DSPace pipeline are grouped at compilation time exploiting a specific bit in the opcodes. In this way, each IP may generate from one to eight sets of instructions executed in parallel, called execution packets (EPs). NOP instructions are automatically dispatched in the decode phase to the idle computational units during the current EP, reducing the memory occupation of the program. At every cycle, if the previous IP has been completely dispatched, a new one is addressed to enter in the pipeline; otherwise, the fetching is stalled. The assignment among instructions and execution units is made explicit into opcode at compile time; thus, the hardware for the dispatch function is minimal while the compiler is rather complex.

The instruction set contains 150 instructions: 92 assigned to the FP_ALUs, 24 to the FP_MUL, and 34 to the AGU. All the instructions can be executed conditionally, depending on the content of some registers in the register files for a reduction of the pipeline hazard.

Three kinds of interrupts are supported in the DPU architecture, served with different levels of priority: reset, nonmaskerable, and seven maskerable interrupts. The source of interrupts may be on-chip or off-chip. The I/O space of the DPU consists of 32 virtual registers that map the peripherals present in the SoC. Accessing these registers with two specific instructions in charge of the FP_ALU results, respectively, in output or input operations involving the mapped peripheral.

Finally, the DPU includes 20 control and status registers grouped in a dedicated register file. There are three main classes of registers: related to the status of the computation of arithmetical, logical, and field operations (i.e., carry, saturation, overflow, and so on); related to the interrupts management (i.e., masking flag, address of handlers table, and so on); and for the configuration of the addressing mode and other DPU settings. FP_ALU can access these registers with a specific instruction.

To improve the system performance while keeping the design simple, two separated single level caches are used for instructions (I-cache) and data (D-cache). High bandwidth and low latency are crucial for these elements. To make these memories space ready, EDAC protection to correct SEC/DED is implemented. Because the cache miss condition stops the DPU fetching until the end of a DMA transfer, the caches are large enough to accommodate entire program functions and data sets typical of matrix-based processing algorithms. For example, 2D elaboration of radar images, where all pixels are iterated in loop way, benefits from caches.

The D-cache is a direct-mapped memory (64 KB, 32 bytes of block size) with copy-back strategy for the cache miss condition. It provides to the AGU units two read ports and two write ports working in parallel for read-read, read-write, write-read, and write-write operations: 64-bit reading without any masking signal (i.e., always 64 bits are returned from the cache) and 32-bit writing operations with byte enable signal are supported.

The actual size of the I-cache is 32 KB, and the block size is 256 bytes. It is implemented in direct-mapped fashion and contains up to eight kilo-instructions of the program. A single 256-bit port enables the DPU to fetch the instruction-packets.

A multichannel DMA engine allows the transfer of data between the caches and the memory controller (off-chip memory). Moving the control of this data path outside the caches gives the programmer the ability to directly manage the local data prefetch/save according to a cacheless paradigm. In fact the DMA controller is mapped in the I/O space of the DPU, and the programmer has the ability to start DMA transfers activating the controller via software instructions.

The memory controller module is the DSPace interface toward the off-chip main memory containing the overall program and the complete set of data to be processed. This is a general module interfacing the DPU with its memory space, enabling transfers from and to the external memory (i.e., EEPROM/SRAM/DRAM/DDRAM).

Finally, the SoC provides two SpaceWire interfaces for high-speed communication with the external hosting system. Using these ports, the DPU can interact with the operative environment receiving data (i.e., samples acquisition from a system sensor) or sending the results of a computation toward other part of the system.

The on-chip bus hierarchy is an AMBA AHB one. The AMBA AHB multimaster bus allows for high-performance communication among the host system, the memory controller, and the control logic. It connects the SpaceWire interfaces (the external environment) with the control logic and memory controller to perform programming and control actions.

The AMBA APB bus enables the DPU (master) to interact directly with the on-chip peripherals (slaves) that represent its own I/O space: DMA, control logic, SpaceWire interfaces. All the peripherals present some memory-mapped registers accessible via the APB bus and dedicated to the configuration and the I/O operations.

The internal memory bus is in charge of moving data between the DMA controller, and the memory controller toward the external main memory. Both instructions and data extracted from the main memory travel on this bus. Indeed, the instruction bus and the data bus are connected, respectively, with the I-cache and D-cache controllers to move instructions and data among the caches and the DMA in case of miss or programmatic transfers.

The DSPace development activity is characterized also by innovation at the methodology design level. Indeed the design and implementation of a new hardware architecture is a challenging task. It takes a long period of development as well as requiring a complex testing phase. The development time becomes longer considering also the tools for the code generation and the system simulator. In fact the SDE design effort is often larger than the hardware design, and it is an error-prone task. A mixed design flow was applied in the DSPace project. The realization of the DPU module went through a methodology of concurrent design of the hardware platform and the related software tools, while the rest of the SoC building blocks were developed following the traditional hardware design flow (i.e., VHDL). Indeed, the DPU was modeled using language for instruction set architectures (LISA) and the associated Processor Designer by Synopsys. LISA is an architecture description language (ADL) that allows the describing of a processor's behavior as well as its structure and instruction set on a very high abstraction level. Using the LISA model, is possible to automatically obtain both the synthesizable VHDL description of the processor and the software development tools (assembler, linker, simulator, debugger). The VHDL model generated by Processor Designer was refined at RTL level to optimize circuit complexity and speed performance of the hardware implementation. Finally all the peripheral blocks were added to the optimized model of the DPU to form the complete model of DSPace SoC. This methodology allows for consistency between the generated software tools and generated hardware, and dramatically improves the quality of the developed processor, reducing the design time.

To complete the SDE, avoiding the cost of the C compiler development, a large reuse of an already existing and stable SDE was adopted. A specific software layer performs code translation and architecture-specific optimizations of the code produced by the reused tool-chain to adapt it to the DSPace architecture and to leverage its parallelism.

The set of software benchmarks representative of typical DSP applications in aerospace, most of them common to radar signal processing, established by the ESA provides a concrete set of criteria for the evaluation of the processor. During the design phase, they contributed to critical decisions while the final validation will provide the results of the DSP performance assessment.

Preliminary synthesis and fitting steps were executed to evaluate the effective resource demand and the operating frequency of the SoC design from the FPGA and ASIC point of view. Also, the demonstration board was addressed.

Targeting as space reference the qualified Virtex5-XQR5VFX130, due to the large demand of silicon and because the SoC saturates the resources of the correspondent ground Virtex5-XC5VLX110, to realize the demonstrator board for the validation, a larger Kintex7-XC7K325T is involved. The complete SoC occupies 30% of the resource budget on this device, operating at 60 MHz of clock frequency. The PCB of the demonstrator has Compact PCI 3U size and includes a DDR2 SODIMM connector with 2 GB of onboard memory. The board was designed in order to be housed in a Compact PCI crate or alternatively in a stand-alone box. Additional SpaceWire, CAN, and MIL1553 communication interfaces are included.

First synthesis results obtained using Synopsys Design Compiler with a CMOS standard cell 180-nm library show an area of around 380 kgates and a peak performance of 1000 MIPS (750 MFLOPS) at 125 MHz. The ATMEL ATC18RHA (standard cell 180 nm) is the target European ASIC technology, but also the ST DSM65 65nm technology will be considered to reach multi-GFLOPS capability.

Preliminary and encouraging results on FIR filtering benchmark, evaluated for different filter lengths (nh = 16, 64, 128, 256, 512, 1024 bit) and various output lengths (nr = 128, 256, 512, 1024 bit), show a performance similar to the fast ground DSP TMSC67XX by Texas Instrument in terms of number of cycles to obtain the result.

3.4 FPGA for Radar Signal Processing

3.4.1 Overview of FPGAs

FPGAs are evolving from reconfigurable hardware devices to complex field programmable systems on chip (FPSoC) with integrated reconfigurable logic blocks (based on FlipFlop and look up tables for the combinatorial logic), I/O blocks with high-speed transceivers (with sustained bandwidth up to Gbps), memory blocks, DSP-dedicated units (multiply, accumulator, adder/subtractor), soft or hard programmable cores (Microblaze, Nios, ARM Cortex) for software programmability.

Therefore, an FPGA can be both software programmable and hardware reconfigurable.

The design of an electronic system on FPGA will require knowledge of both hardware-oriented languages and tools (e.g., hardware description languages like VHDL or Verilog) and of software-oriented languages and tools (e.g., C or assembly for programmable core programming, operating systems).

Therefore, managing an FPGA technology can be a bit harder than managing a pure software programmable platform, although there are several tools, libraries of predesigned components, and languages whose aim is allowing the design of FPGA-based systems from high abstraction levels in C or Matlab Simulink or in graphic environments (CatapulC, DSPbuilder, System Generator for DSP)

Different FPGA technologies exist from many vendors, the leaders being Altera, Xilinx, Actel (now Microsemi). For example, flash technology is adopted in the ProAsic family, antifuse technology in AX and RTAX families, SRAM in Virtex, Kintex, Spartan, Arria, Cyclone, and Stratix devices.

For computing-intensive applications like radar signal processing and considering that the target of this book is civil applications, SRAM-based FPGAs are typically preferred.

Mixed-signal FPGAs also exist, like the fusion family integrating in the FPGA digital hardware configurable blocks, software programmable core (e.g., a Cortex-M1 from ARM), and analog blocks like an ADC and analog quads to scale up or down signals.

Some FPGA families are already qualified for automotive or space/military applications. Some vendors, like Altera or Xilinx, have dedicated developments for radar signal processing application. With respect to an ASIC design, an FPGA design allows for short time to market, fewer design risks, increased flexibility, and lower costs for markets that are still not entered in a large-volume production, which is the case of low-power radar for civil applications.

To be noted that the FPGA design can be seen as a first step toward an ASIC. Indeed, most vendors enable an easy way to migrate from FPGA system design to prototyping and then into volume production. By using one methodology, one tool environment, one set of intellectual property (IP) cores, the designer can design and prototype FPGAs and then ramp production when it makes sense for the market, and the design no longer requires the full programmability of standard FPGAs. This is the case of the hardcopy products of Altera or Easy Path from Xilinx.

The scope of the chapter is not reviewing all recent advances in the field of FPGA from all vendors but showing the capabilities of these devices for radar signal processing applications. This is why, for example, Table 3.6 we summarize the main performance of the FPGA families of just one vendor, Xilinx. Other vendors have product lines with comparable performances.

Current Xilinx FPGA devices include SRAM-based families realized in different technology nodes and addressing different trade-offs between performance and cost/complexity: from the Spartan family realized in 45-nm technology node, then there are the Artix, Kintex, and Virtex families in 28-nm technology node, and the emerging ultrascale Virtex and Kintex devices in 20-nm technology and lower. The lower the technology node scale, the higher the complexity and functionality of the FPGA but also the cost.

As we can see from the Table 3.6, today FPGA devices integrate RAM storage resources on the order of a few to several tens of Mb, with high-speed dedicated interfaces toward off-chip DRAM.

Modern FPGAs also integrate from few hundred to several thousands of DSP blocks with multiply and accumulate (MAC) capabilities. The overall GMAC/s performances are up to tera multiply and accumulate operations per second, and hence FPGA can compete with the most complex GPU platforms.

Table 3.6
Xilinx FPGA Comparison Table

	Spartan-6	Artix-7	Kintex-7	Virtex-7
Logic cells	150,000	215,000	480,000	2,000,000
BlockRAM	4.8 Mb	13 Mb	34 Mb	68 Mb
DSP slices	180	740	1,920	3,600
DSP performance (symmetric FIR)	140 GMACs	930 GMACs	2345 GMACs	5335 GMACs
Transceiver count	8	16	32	96
Transceiver speed	3.2 Gbps	6.6 Gbps	12.5 Gbps	28.05 Gbps
Total transceiver bandwidth (full duplex)	50 Gbps	211 Gbps	800 Gbps	2,784 Gbps
Memory interface (DDR3)	800 Mbps	1066 Mbps	1866 Mbps	1866 Mbps
PCI Express® interface	x1 Gen1	x4 Gen2	x8 Gen2	x8 Gen3
I/O pins	576	500	500	1,200
I/O voltage	1.2V, 1.5V, 1.8V, 2.5V, 3.3V	1.2V, 1.35V, 1.5V, 1.8V, 2.5V, 3.3V	1.2V, 1.35V, 1.5V, 1.8V, 2.5V, 3.3V	1.2V, 1.35V, 1.5V, 1.8V, 2.5V, 3.3V

Multiple I/O transceivers, running from few to tens of Gbps, are also available in FPGA devices. Even PCI express interfaces can be available.

3.4.2 High-End FPGA for Radar Signal Processing

The overall complexity of high-end FPGAs is on the order of billions of transistors, occupying an area of several hundreds of mm2, and the power consumption is up to tens of Watts.

For example, a device like the Virtex6 LX760 FPGA, belonging to the previous Virtex generation (realized like the Spartan family in 45-nm technology), offers a DSP computational capability of roughly 380 GFLOPS when implementing a FFT algorithm but for a power cost higher than 50W.

If we compare the performance and power cost of this FPGA to an ASIC realization of the same FFT algorithm or to a software implementation on a general purpose processor like an INTEL CoreI7 or a GPU, (see Figure 3.8), like the GTX285 and GTX480 platforms, the following considerations can be done:

- The LX760 FPGA has worse performances, measured as the ratio between computational capability and power consumption (GFLOPS/W) or area (GFLOPS/mm2), by one order of magnitude versus a dedicated ASIC design.
- The LX760 FPGA has similar performances to a GPU (GTX480 from NVidia) realized in the same 45-nm CMOS technology and one order of magnitude better performance than the general purpose processor.

It should be noted that in Figure 3.8, the area is normalized versus a single technology node at 40 nm taken as reference with a factor λ_2/λ_1 being λ_2 the channel length of the MOS devices in the original technology and $\lambda_1 = 40$ nm.

Moreover, for the FFT different lengths are considered. For FFTs, the computation load increases as $N \log_2 N$, whereas the data I/O increases as N.

It is worth noting that the shorter FFT lengths are prevalent in radar signal processing, where FFT lengths of 512 to 8192 are the norm.

Therefore, high-end FPGAs for intensive radar signal processing are available but their cost (> US$1000) and power consumption are too high for ubiquitous low-power applications. Rather, these FPGAs are more suited for defense, avionic, or marine radar applications.

For example, a synthetic aperture beam forming radar design for defense applications has been proposed by Xilinx based on Kinetx-7 devices. The device has two 10G Ethernet and eight PCIe interfaces versus DRAM and a host CPU. The following radar signal processing tasks for 48 TX and RX channels

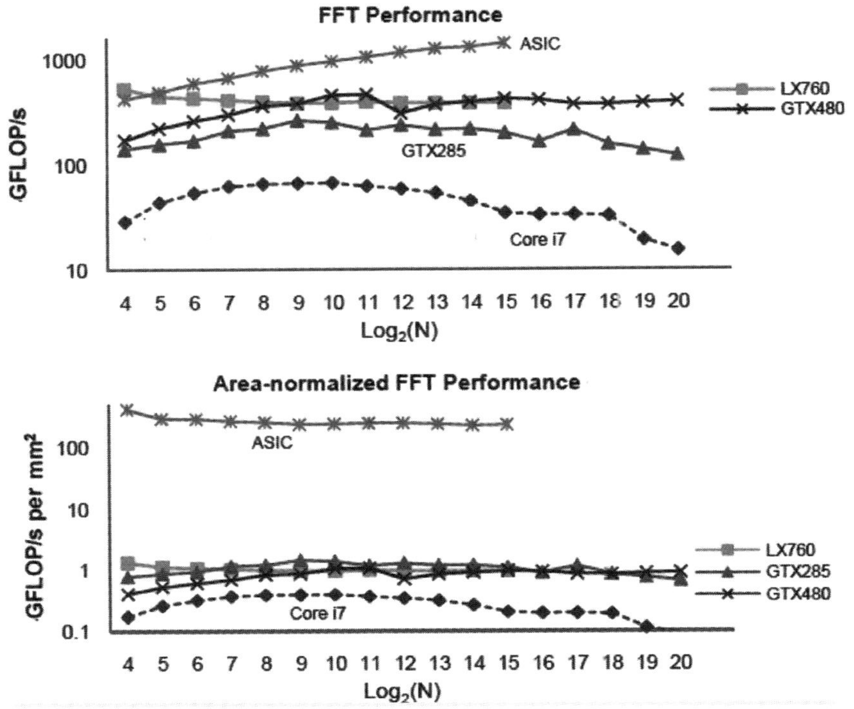

Figure 3.8 Comparison of a Virtex6 LX760 FPGA versus GPP, GPU, and ASIC for FFT processing. The analysis confirms results that are known in the literature (e.g., [38]).

are integrated in the FPGA resources: digital-down conversion, RX phase shifter, RX adaptive beamformer, digital up-conversion, TX waveform generation, TX direct digital synthesis, and pulse compression.

Although with respect to a previous radar generation, still for defense applications, this FPGA approach claims for the realization of a single-board radar with a reduction of 50% in device use, 30% in bill of material cost, and 50% in power consumption, the power consumption and cost are still prohibitive for low-power/low-cost civil applications.

Similar considerations can be done for a single board design with three Virtex™-7 980T FPGAs proposed by Xilinx and implementing a 64-channel beam-former radar.

High-end FPGA designs for radar signal processing have also been proposed by Altera.

For example, in [6] a Stratix Altera FPGA design in 28 nm allows for variable precision DSP blocks (multipliers 18×18 or 27×27or 18×36, and so on, plus 64-bit accumulation) with a capability of tera multiply and accumulate operations per second. Several radar DSP functions have been implemented using Stratix FPGAs such as STAP, digital beam-forming, and pulse compression [28–31].

In a recent Altera white paper [29] dated May 2013, "Radar Processing: FPGAs or GPUs?" Altera's next-generation high-performance FPGAs was announced to support a minimum of 5 TFLOPs performance by leveraging Intel's 14-nm Tri-Gate process. Up to 100 GFLOPs/W can be expected in the future using this advanced semiconductor process.

The current state of the art is represented by devices like the 28-nm Stratix V 5SGSD8 FPGA that in the same white paper is demonstrated having the capability to support in real time 4096-point Radix2 FFT.

The GFLOPS performance of such a device is from 81 to more than 500 depending on the level of optimization of the design; in terms of power efficiency, a figure of 10 GFLOPS/W can be achieved with StratixV FPGA. Obviously such performance figures depend also on the adopted algorithms for the benchmarks. For example, using a more complex algorithm like STAP with Cholesky decomposition, the achieved efficiency is about 5 GFLOPs/W [32].

3.4.3 Cost-Effective FPGA for Radar Signal Processing

The analysis reported in the previous paragraph clearly demonstrates the signal processing capabilities of FPGAs [33–40], which can compete with the best-performing GPU with computational capabilities on the order of tera floating-point operations per second.

However, our target applications of a portable radar for vital sign detection or a radar unit for automotive market require fitting the DSP radar requirements on lower-cost lower-power platforms [30, 33–35].

Recent results in the literature, from both academia and industry, prove that this can be achieved through an optimized algorithmic-architecture co-design on less complex FPGA families with hardware-software capabilities.

For example, in the works by Lal [34, 35], the microcontroller and DSP functionalities for a 77-GHz FMCW automotive radar are realized using a Xilinx Virtex5 SX50T device.

The radar front-end at radio frequency and intermediate frequency has been realized with GaAs MMIC technology.

Control and baseband DSP functionalities are carried out by a Virtex-5 SX50T, which manages switch control, TX sweep generation, Hamming filtering, FFT (radix-2 2048-point), CA-CFAR, and peak pairing. The ADC and DAC are sized with 11b and 10b resolution, respectively.

The Virtex5 SX50T devices has 8160 slices (4 LUT + 4FF), 5 Mb RAM, 288 DSP48 unit (25x18 multiply plus accumulate plus adder), 12 digital clock management, and fast I/O. The clock frequency is sized at 100 MHz.

A characteristic of this radar is that through a microelectro mechanical system (MEMS) controllable switch, the antenna array can be configured so that the devices can work as SRR, medium-range radar (MRR), or long-range radar (LRR).

In the context of this book (see also Table 3.7), the term SRR refers to a radar with a detection range in the order of few meters, with a maximum 30m; the term MRR refers to a radar with a detection range in the order of tens of meters, with a maximum 100m; and the term LRR refers to a radar with a detection range in the order of hundreds of meters, with maximum 200–250m. In Table 3.7, the achieved performance of this FPGA-based automotive radar is compared to those of a commercial product: the third generation long-range radar from BOSCH (LRR3).

Other designs reported in the literature for FMCW radar proposed the integration on a Virtex2PRO FPGA of a Microblaze programmable core plus, fitted on the logic resources of the FPGA, a coprocessor realizing a 2048-point radix-2 FFT with 14-bit resolution.

Industrial examples of automotive radars whose signal processing chain is realized on a low-cost FPGA family, the Cyclone, for which automotive qualification is available, have been recently proposed by Altera [30].

The proposed designs are based on a family of devices we called FPSoC. Indeed, the digital device embeds two 600-MHz ARM Cortex-A9 processors beside the standard FPGA resources. Automotive-grade versions of these devices are also available and capable of 600-MHz CPU clock rates.

The designs from Altera include both a pulse-Doppler radar and an FMCW radar.

The pulse-Doppler radar sends successive pulses at specific intervals or a pulse repetition interval (PRI). As the radar returns, the reflections are processed coherently to extract range and relative motion of detected objects.

Table 3.7
Performance of the FMCW Automotive Radar Based on Virtex 5 FPGA from [34] versus Commercial LRR3 Product from Bosch

		MEMS Tri-Mode (Xilinx Virtex 5 HDL Simulation)		
Parameter	**Bosch LLR3**	**SRR**	**MRR**	**LRR**
Range (m)	0.5–250	0.4–30	30–100	100–200
Velocity (km/h)	−100 to +200	±300	±300	±300
Range accuracy (m)	±0.10	±0.10	±0.14	±0.28
Velocity accuracy (m/s)	±0.12	±0.14	±0.42	±0.83
Processing latency	N/A	106 μs	212 μs	212 μs
Cycle time	80 ms	62 ms for three modes combined		

In automotive applications, the range can be as short as a few meters to as much as a few hundred meters. For a range of 2m, the round-trip transit time of the radar pulse is 13 ns. Achieving a 2m range resolution requires a sample rate in the order of 100 mega samples per second (MS/s) or more.

Table 3.8 reports the implementation complexity results of a DSP chain for pulse-Doppler radar with PRI of 250 KHz, 64 range bins, and 65 Doppler bins realized on a Cyclone V FPGA, 5CSEA5 device.

The design occupies 42% of the available FPGA resources (Logic Elements LE), 76% of the available dedicated DSP blocks, and 40% of the available memory resources. The maximum achievable frequency for the FPGA clock rate is roughly 200 MHz.

It should be noted that in the Altera realization the DSP units are capable of variable precision computation de facto emulating a floating-point arithmetic.

Altera has also reported an example of the implementation with a Cyclone V FPGA of the processing chain of a 77-GHz FMCW automotive radar.

Figures 3.9 and 3.10 report, respectively, the analog section and the baseband digital section of the implemented 77-GHz FMCW automotive radar.

On the analog side, the transmitter waveform can be implemented using a direct digital synthesizer (DDS) with a standard reference crystal. The DDS generates an analog frequency ramp reference for the phased-locked loop (PLL) to generate the desired transmit frequency modulation. For example, if the PLL has a divider of 1000, then the reference would be centered at 77 MHz. This analog ramp signal drives the reference of a PLL, which disciplines a 77-GHz oscillator. The oscillator output of the circuit is amplified and produces the continuous wave (CW) signal ramping up and down over 500 MHz with a center frequency of 77 GHz. Filtering and matching circuits at 77 GHz are accomplished using passive components etched into high-epsilon dielectric circuit cards, minimizing the components required.

In the receiver, the front end requires filtering and a low-noise amplifier (LNA), followed by an analog mixer. The mixer down-converts the 77-GHz receiver signal with the ramping transmit signal, outputting a baseband signal that

Table 3.8

Complexity of a DSP Vhain of a Pulse-Doppler Radar Implemented on a Cyclone V FPGA 5CSEA25 Device (64 Ranges Bins, 64 Doppler Bins, 100-Mhz Sample Rate, PRI 250 kHz)

	Logic Elements	Variable-Precisions DSP Blocks	Memory Blocks	Max Clock Frequency
Cyclone5 5CSEA5 available resources	85K	87	397	197 MHz at 1.1V and 85C
Percent FPGA occupation	42%	76%	39%	

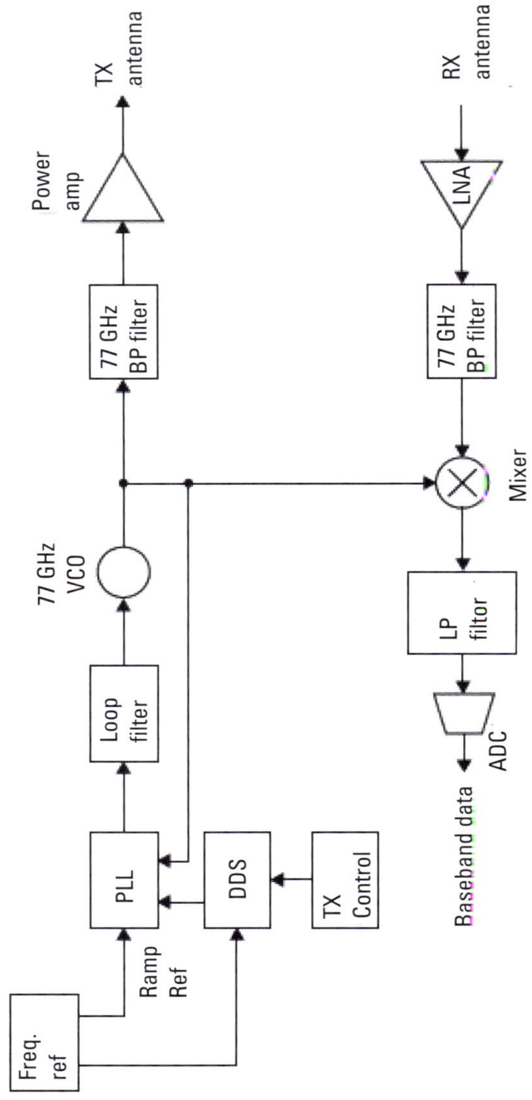

Figure 3.9 FMCW Automotive radar scheme similar to that proposed by Altera on Cyclone V—analog section [30].

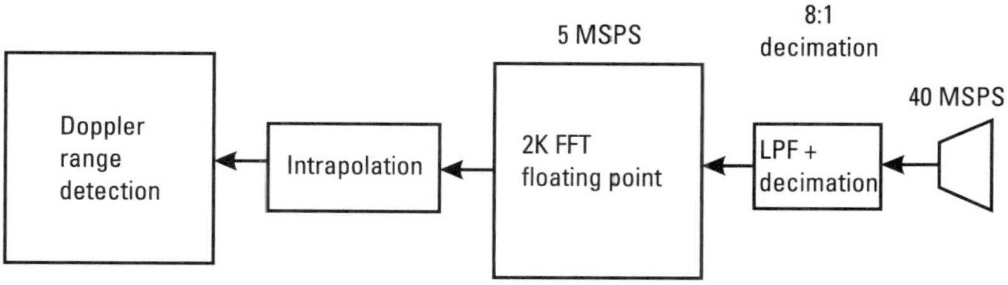

Figure 3.10 FMCW automotive radar scheme similar to that proposed by Altera on Cyclone V—digital section [30].

contains the difference between the transmitter and receiver waveforms at any given instant. The baseband analog content has a bandwidth up to few MHz. The high-frequency filtering at 77 GHz can be implemented using etched passive components. The output of the mixer will be at low frequency (around 2 MHz). Therefore, traditional passive components and operational amplifiers can be used to provide antialiasing low-pass filtering prior to the ADC.

The baseband signal is composed of frequencies, either all positive (during negative frequency ramp) or all negative (during positive frequency ramp), so a mixer followed by a single low-pass filter and ADC is sufficient.

The ADC for baseband input must operate at a minimum of 5 mega sample per second (MSPS) to meet the Nyquist criterion. If, instead, an 8x sampling frequency of 40 MSPS is used, followed by an 8:1 digital decimation filter, then approximately three bits of additional resolution can be achieved. This decimation allows a 16-bit ADC to effectively operate in the range of 18 to19 bits, achieving well over 100 dB of dynamic range.

The next step of signal processing is to perform frequency discrimination using an FFT, followed by an interpolation circuit.

The solution proposed by Altera implements a 2048-point FFT, in single-precision floating-point processing, and is claimed to support the 100-dB dynamic of the target returns. Distant and weak target returns will not be obscured by nearby, strong returns, therefore avoiding the radar system being "blinded" by strong, nearby reflections. The floating-point processing will also prevent weak returns from being "buried" in the digital noise floor of the FFT.

The floating-point FFT circuit in the Altera solution is also clocked at 160 MHz, though only processing complex data at 10 MSPS. All this chain is fitted on the low-cost Cyclone V SoC FPGAs, which feature DSP blocks that support both 18×19 and 27×27 multiplier sizes. The larger 27×27 multiplier size is ideal for use in floating-point designs.

An interpolation of the FFT output will help with precise frequency discrimination. If their energy is between frequency bins, the interpolation will prevent weak targets from being missed by aliasing the FFT energy into other bins. Detection processing will likely be done using one of the on-chip ARM

Cortex-A9 microprocessors. Algorithms from simple threshold to more sophisticated CFAR can be used. The interpolation of the FFT results can be implemented either in logic or in software, depending on the system designer's choices.

Table 3.9 reports the implementation complexity results of an FMCW automotive radar on the Cyclone V FPGA, 5CSEA2 device.

The design occupies 24% of the available FPGA resources (Logic Elements LE), 22% of the available dedicated DSP blocks and 21% of the available memory resources. The maximum achievable frequency is roughly 190 MHz.

In comparing Tables 3.8 and 3.9, it is clear that the FMCW design uses much fewer resources than the pulse Doppler design, and hence the FMCW radar can be implemented in cost-effective FPGA for automotive market such as the Cyclone5 SEA2.

Table 3.10 summarizes the main characteristics of CycloneV devices.

3.5 Conclusions

The aim of this chapter was reviewing possible implementing platforms for radar digital signal processing, with a particular focus on those meeting the requirements in terms of computational capability, flexibility, and low power consumption of the target ubiquitous civil radar applications.

While in the origin, a radar was mainly an analog system; now it is becoming a mixed-signal system. Radar signal can be digitized directly at intermediate frequency, minimizing the daunting tasks to be carried out in the analog domain at radio frequency. Fast Fourier transform and its inverse, time domain and frequency domain filtering, pulse compression, digital down- and up-conversion, beam-forming, waveform generation by direct digital synthesis,

Table 3.9
Implementation Results of a FMCW Radar in Cyclone V FPGA 5CSEA2 Device

	Logic Elements	Variable-Precisions DSP Blocks	Memory Blocks	Max Frequency
Fixed-point decimation filter	1 K	4	0	190 MHz, At 1.1 V, 85 C
Floating-point 2K FFT	5 K	4	29	
Total resources	6 K	8	29	
Cyclone5 5CSEA2 available resources	25 K	36	140	
Percent FPGA occupation	24%	22%	21%	

Table 3.10
Cyclone V SE FPSoC for Cost-Effective Radar Signal Processing Implementation

Device	5CSEA2	5CSEA4	5CSEA5	5CSEA6
LEs (K)	25	40	85	110
M10K memory blocks	140	224	397	514
M10K memory (Kb)	1400	2240	3972	5140
18-bit x 19-bit multipliers	72	116	174	224
Variable-precision DSP blocks	36	58	87	112
FPGA PLLs	4	5	6	6
Package	U484, U 672	U484, U 672	U484, U 672, F896	U484 , U 672, F896
Maximum FPGA user I/Os	145	145	288	288
FPGA hard memory controllers	1	1	1	1
Processor cores (ARM Cortex™-A9 MPCore™)	Single or dual	Single or dual	Single or dual	Single or dual

and CFAR are some examples of common radar signal and data processing algorithms to be realized in the digital domain. Beside processing functions, the radar digital domain should be in charge of control and interface tasks. For example, in the automotive domain the radar subsystem should be connected to the main control vehicle network through a CAN bus interface.

Lots of possible implementation platforms are available in the literature with different trade-offs between computational and memory capabilities, power consumption and area, development time and costs, and flexibility: from hardware-oriented solutions (ASIC, FPGA) to software-based ones (MCU, DSP, GPU, GPP) to mixed hardware-software platforms (FPSoC, MPSoC).

The market of ubiquitous radar for civil applications is still not a mature large-volume market, with frozen standards and reference architecture, and hence an ASIC approach is not suitable. GPP can offer the required computational capabilities for radar signal processing and maximize system flexibility, but its power consumption is too high. Higher efficiency in terms of computational capability provided versus the power consumption, while still keeping a software-oriented approach, can be offered by high performance DSP, like many-core DSP processors (e.g., CELL) or GPU. However, the absolute values of computational power, up to TFLOPS, and power, much higher than 1W, can be suited for radar applications in defense or surveillance applications (avionic, marine) but not for portable ubiquitous civil radar as those discussed in the next chapters. Similar considerations are still valid for high-end FPGA devices with too-high computational capabilities, power, and device cost.

For the target application of this book, the most suited platform implementation for the baseband digital domain is represented by a MCU core, in charge of basic control and networking functions, plus a low-power/low-cost FPGA or DSP coprocessors for computing intensive signal processing operators. The functionalities of the MCU core and the FPGA or DSP coprocessors can be now integrated in a single device class of the FPSoC device that are FPGAs with embedded hard or soft programmable cores. This approach can provide a cost-effective single-board solution with reduced development time and cost. Moreover, in case of future market success, an easy migration for an ASIC implementation, with reduced cost in a large volume market, is available.

As discussed, this proposed approach is not only a vision but, since 2013 the market has already been offering components to implement it.

References

[1] Fleming, B., "Recent Advancement in Automotive Radar Systems," *IEEE Vehicular Technology*, Vol. 7, No. 1, 2012, pp. 4–9.

[2] Menxel, W., et al., "Antenna Concepts for Millimeter-Wave Automotive Radar Sensors," *Proceedings of the IEEE*, Vol. 100, No. 7, 2012, pp. 2372–2379.

[3] Neri, B., and S. Saponara, "Advances in Technologies, Architectures and Applications of Highly Integrated Low-Power Radars," *IEEE Aerosp. Eelectr. Syst.*, Vol. 27, No. 1, 2012, pp. 25–36.

[4] Stevenson, R., "A Driver's Sixth Sense," *IEEE Spectrum 2011*, Vol. 48, No. 10, pp. 50–55.

[5] Winkler, V., "Range Doppler Detection for Automotive FMCW Radars," *Proc. European Radar Conference*, 2007, pp. 166–169.

[6] Mauer, V., and M. Parker, "Floating-point STAP Implementation on FPGAs," *Proc. IEEE Radar Conference*, Kansas City, KS, 2011, pp. 901–904.

[7] Schoor, M., and B. Yang, "High-Resolution Angle Estimation for an Automotive FMCW Radar Sensor," in *Proc. IEEE International Radar Symposium (IRS)*, Cologne, Germany, Sept. 2007.

[8] Wenig, P., et al., "System Design of a 77-GHz Automotive Radar Sensor with Super Resolution DOA Estimation," *Proc. International Symposium on Signals, Systems and Electronics*, 2007, pp. 537–540.

[9] Kok, D., and J. S. Fu, "Signal Processing for Automotive Radar," *Proc. IEEE International Radar Conference*, Arlington, VA, May 9–12, 2005, pp. 842–846.

[10] Whil, R., et al., "Fast Growing, Diversified Automotive Radar Market Presents Chip Vendors with New Challenges," *Proc. of IWPC (International Wireless Packaging Consortium) workshop*, 2011, p. 1–20, http://automotive.analog.com/static/imported-files/tech_docs/adi-2011.pdf.

[11] Trotta, S., et al., "An RCP Packaged Transceiver Chipset for Automotive LRR and SRR Systems in SiGe BiCMOS Technology," *IEEE Transaction on Microwave Theory and Technique,* Vol. 60, No. 3, 2012, pp. 778–794.

[12] Kawano, Y., et al., "A 77-GHz Transceiver in 90-nm CMOS," *Proc. IEEE Int. Conf. on Solid State Circuits (ISSCC),* San Francisco, CA, Feb. 8–12, 2009, pp. 310–311.

[13] Li, Y. A., et al., "A Fully Integrated 77-GHz FMCW Radar Transciever in 65-nm CMOS, Technology," *IEEE Journal Solid-State Circuits,* Vol. 45, No. 12, 2010, pp. 2746–2756.

[14] Mitomo, T., et al., "A 77-GHz 90-nm CMOS Transceiver for FMCW Radar Applications," *IEEE Journal Solid-State Circuits,* Vol. 45, No. 4, 2010, pp. 928–937.

[15] Richter, M. D. et al., "77-GHz Automotive Digital Beam-Forming Radar with SiGe chipset," *Proc. German Microwave Conference,* 2010, pp. 210–213.

[16] Hasch, J., et al., "Millimeter-Wave Technology for Automotive Radar Sensors in the 77-GHz Frequency Band," *IEEE Transaction on Microwave Theory and Technique,* Vol. 60, No. 3, 2012, pp. 845–860.

[17] Pam, D. C., et al., "Overview of the Architecture, Circuit Design, and Physical Implementation of a First-Generation Cell Processor," *IEEE Journal of Solid State Circuits,* Vol. 41, No. 1, 2006, pp. 179–193.

[18] Parakh, N., A. Mittal, and R. Niyogi, "Optimization of MPEG 2 Encoder on Cell B. E. Processor," *IEEE Int. Advanced Computing Conference,* 2009, pp. 423–427.

[19] Pille, J., et al., "Implementation of the Cell Broadband Engine™ in 65-nm SOI Technology Featuring Dual Power Supply SRAM Arrays Supporting 6 GHz at 1.3 V," *IEEE Journal of Solid State Circuits,* Vol. 43, No. 1, 2008, pp. 163–171.

[20] Takahashi, O., et al., "Migration of Cell Broadband Engine from 65 nm SOI to 45 nm SOI," *Proc. IEEE Int. Solid State Circuit Conf. (ISSCC),* Feb. 2008, San Francisco, CA, pp. 85–87.

[21] Chung, E., et al., "Single-Chip Heterogeneous Computing: Does the Future Include Custom Logic, FPGAs, and GPGPUs?," *Proc. IEEE/ACM 43rd Micro Conference,* 2010, pp. 223–225.

[22] Ma, X., et al., "Statistical Power Consumption Analysis and Modeling for GPU-Based Computing," *Proc. Workshop on Power Aware Computing and Systems,* Big Sky, MT, Oct. 2009, pp. 1–5.

[23] Nickolls, J., et al., "The GPU Computing Era," *IEEE Micro,* Vol. 30, No. 2, 2010, pp. 56–69.

[24] Pettersson, J., and I. Wainwright, "Radar Signal Processing with GPU," SAAB Doc. n. 5/0363-FCP1041180 en, 2010, p. 127.

[25] Trautner, R., "ESA's Roadmap for Next Generation Payload Data Processors," *Proc. of DASIA Conference,* Malta, May 2011, pp. 1–5.

[26] Baronti, F., et al., "Design and Verification of Hardware Building Blocks for High-Speed and Fault-Tolerant In-Vehicle Networks," *IEEE Transactions on Industrial Electronics,* Vol. 58, No. 3, 2011, pp. 792–801.

[27] Errico, W., et al., "DSPace: A New Space DSP Development," *Proc. of DASIA Conference, Dubrovnik, Croatia, May 14–16, 2012,* pp. 706–711.

[28] Parker, M., "Achieving One TeraFLOPS with 28-nm FPGAs," *Electronic Engineering Journal,* doc. n. WP-01142, Sept. 2010, pp. 1–13.

[29] Parker, M., "Radar Processing: FPGAs or GPUs?" *Electronic Engineering Journal,* doc. n. WP-01197, rev. 2.0, 2013, pp. 1–11.

[30] Parker, M., "Implementing Digital Processing for Automotive Radar Using SoC FPGAs," *EEtimes,* Nov. 2012, http://www.eetimes.com/document.asp?doc_id=1280099.

[31] Land, I., "Floating-Point FPGAs for DSP Bring High Precision to Radar and EW Systems," *ECN (Electronic Components News),* No. 4, 2011, http://www.ecnmag.com/articles/2011/04/floating-point-fpgas-dsp-bring-high-precision-radar-and-ew-systems.

[32] Yang, D., et al., "Performance Comparison of Cholesky Decomposition on GPUs and FPGAs," *Proc. Symposium on Application Accelerators in High Performance Computing (SAAHPC),* Knoxville, TN, 2010, pp. 1–3.

[33] Saad, J., et al., "FPGA-Based Radar Signal Processing for Automotive Driver Assistance System," *Proc. IEEE Int. Symp. on Rapid System Prototyping (RSP),* Paris, France, 2009, pp. 196–199.

[34] Lal, S., "An FPGA-Based Signal Processing System for a 77-GHz MEMS Tri-Mode Automotive Radar," *Proc. IEEE International Symposium on Rapid System Prototyping (RSP) 2011,* Karlsruhe, Germany, pp. 2–8.

[35] Lal, S., et al., "An FPGA-Based 77 GHZ MEMS Radar Signal Processing System for Automotive Collision Avoidance," *Proc. IEEE Canadian Conference on Electrical and Computer Engineering (CCECE),* 2011, Niagara Falls, NY, pp. 1351–1356.

[36] Parker, M., "Radar Basics Tutorial Part 1," *EEtimes,* May 2011, http://www.eetimes.com/document.asp?doc_id=1278779.

[37] Parker, M., "Radar Basics Part 2: Pulse-Doppler Radar," *EEtimes,* May 2011, http://www.eetimes.com/document.asp?doc_id=1278808.

[38] Parker, M., "Radar Basics Part 3: Beamforming and Radar Digital Processing," *EEtimes,* June 2011, http://www.eetimes.com/document.asp?doc_id=1278838.

[39] Parker, M., "Radar Basics Part 4: Space-Time Adaptive Processing," *EEtimes,* June 2011, http://www.eetimes.com/document.asp?doc_id=1278878.

[40] Parker, M., "Radar Basics Part 5: Synthetic Apature Radar," *EEtimes,* July 2011, http://www.eetimes.com/document.asp?doc_id=1278931.

4

Radar for E-Health Applications: Signal Processing Perspective

4.1 General Characteristic of the Sensor and Its Functions

Medical technology has improved remarkably over the past 30 years, becoming less invasive and more sophisticated as the years progress. Until recently, however, completely noncontact health monitors have existed only in science fiction. Now, with new technologies, noncontact respiration and heartbeat monitoring offer a real, attractive alternative to commonly prescribed cheststrap monitors.

The noncontact remote detection of vital signs leads to several potential applications, such as monitoring sleeping infants or adults to detect abnormal breathing conditions and searching survivors after earthquakes. Or, very important in aging populations, home monitoring of vital signs, which can decrease the cost of health care by moving some patients from hospital to home.

While the concept of noncontact detection of vital signs has been successfully demonstrated by pioneers in this field before 2000 [1–3], research efforts in the last decade have boosted the technology development toward lower power, lighter weight, longer detection range, with better accuracy and greater robustness for portable and handheld applications. There are several advantages to a noncontact measurement method. It does not inhibit the subject under test or cause discomfort or skin irritation. And this is particularly important over long terms of continuous monitoring.

As for the automotive applications (see Chapter 5), two different types of radar are used for detecting vital signs, in particular for heartbeat and respiration monitoring: CW Doppler radars and UWB radars, which use different principles for detection.

4.2 CW Doppler Radar for Health Care Monitoring

According to Doppler theory, a target with a time-varying position, but an average zero velocity, will reflect the signal with its phase modulated proportionally to the time-varying target position. A stationary person's chest has a periodic movement with an average zero velocity, and, therefore, a CW radar with the chest as the target will receive a signal similar to the transmitted signal with its phase modulated by the time-varying chest position [4].

Then, in CW radar systems, an unmodulated signal $s_T(t) = \cos(2\pi f_0 t + \phi(t))$ with carrier frequency f_0 and phase $\phi(t)$ is transmitted toward a human body at a nominal distance of R_0, where it is phase modulated by the physiological movement (heartbeat and respiration) represented by the signal $x(t)$.

The fundamental mechanism used for detection in CW Doppler radar is illustrated in Figure 4.1.

Neglecting amplitude variation and assuming that $x(t) << R_0$, the reflected signal recorded by the radar receiver, after the round-trip time $t_0 = 2R_0/c$, is given by

$$s_R(t) = \cos\left[2\pi f_0 t - \frac{4\pi R_0}{\lambda} - \frac{4\pi x(t)}{\lambda} + \phi\left(t - \frac{2R_0}{c}\right)\right] \tag{4.1}$$

where c is the speed of light and λ is the radar wavelength. So the received signal is similar to the transmitted one, with a time delay t_0 and a phase modulated by the periodic body movement. The aim of the CW radar is to detect and analyze the signal $x(t)$ that bears the information on the heartbeat and respiration rhythm.

Using the transmitted signal $s_T(t)$ as LO signal, the radar down-converts $s_R(t)$ into the baseband signal

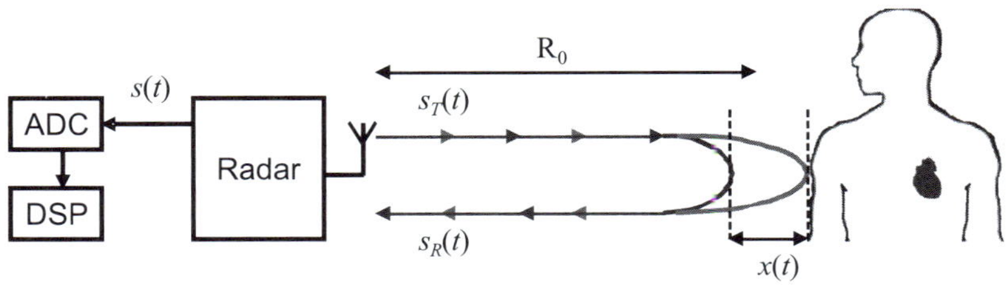

Figure 4.1 Fundamental mechanism of CW Doppler radar vital sign detection.

$$s(t) = \cos\left[\frac{4\pi x(t)}{\lambda} + \Delta\phi(t) + \theta\right] \tag{4.2}$$

where $\theta = \theta_0 + 4\pi R_0/\lambda$ is determined by the nominal distance R_0 and θ_0 is a phase shift and $\Delta\phi(t) = \phi(t) - \phi\left(t - \frac{2R_0}{c}\right)$ is the residual phase. Several factors affect the value of θ_0 such as the phase shift at the reflection surface and any distance between the mixer and the antenna. Phase θ_0 is a fixed or an adjustable shift that is assumed to remain constant during the period of signal processing.

When the movement amplitude is much smaller than the radar wavelength λ, a linear approximation can be applied to (4.2). If θ is an odd multiple of $\pi/2$, the baseband output can be approximate as

$$s(t) \simeq \frac{4\pi x(t)}{\lambda} + \Delta\phi(t) \tag{4.3}$$

In this case, it is evident that $s(t)$ is proportional to the motion displacement signal $x(t)$ summed to the residual phase noise, and the optimum phase-demodulation sensitivity is achieved.

If, conversely, θ is an even multiple of $\pi/2$, the baseband output can be approximate as

$$s(t) \simeq 1 - \left[\frac{4\pi x(t)}{\lambda} + \Delta\phi(t)\right]^2 \tag{4.4}$$

In this case, $s(t)$ is no longer linearly proportional to $x(t)$; the sensitivity is decreased, and *null-points detection* can occur. A null point occurs when the LO and received signal are either in-phase or 180 degrees out of phase. Since the variable part of θ is dependent only on the distance to the target R_0, this null point occurs with a target distance every $\lambda/4$ from the radar. At a frequency of 2.4 GHz, typical of this kind of radar, these null points occur every 3 cm and therefore it can be difficult to avoid them solely by adjusting the position of the transceiver.

In [4, 5] the authors proposed to avoid these null detection points with a quadrature receiver, where two receiver chains with LO phases 90 degrees apart insure that there is always at least one output not in the null detection point. The two in-phase (I) and quadrature (Q) output channels of a quadrature receiver will then be

$$
\begin{aligned}
s_I(t) &= \cos\left[\frac{4\pi x(t)}{\lambda} + \Delta\phi(t) + \theta\right] \\
s_Q(t) &= \sin\left[\frac{4\pi x(t)}{\lambda} + \Delta\phi(t) + \theta\right]
\end{aligned}
\tag{4.5}
$$

When θ is an even/odd multiple of $\pi/2$, the I channel signal will be at the null/optimum detection point while the Q channel signal will be at the optimum/null detection point. In these cases, there will always be one channel at the optimum detection point to guarantee good detection accuracy.

The signal $x(t)$ is due to the movement of the chest caused by both heartbeat and breathing.

Breathing is an involuntary periodic activity in which each period consists of three main parts: inhalation takes about 1s, exhalation takes about 1s, and relaxation takes about 2s. The total amount of air transferred during one period, which is called tidal volume, is 500–600 ml. The respiratory rate is 12–20 breaths per minute for adults [6].

The heart drives blood through the lungs and to tissues throughout the body. When the heart contracts to generate the pressure that drives blood flow, it moves within the chest cavity, hitting the chest wall and creating a measurable displacement at the skin surface. The contraction and relaxation of the left ventricle cause a larger chest motion than do other heart actions in healthy subjects.

In many papers, maximum displacement of chest wall due to heart motion has been reported to be less than 1 mm. Heart rate is the number of heartbeats per unit of time, typically expressed as beats per minute (bpm). The typical resting heart rate in adults is 60–80 bpm [6].

Based on these physical considerations, we can express the signal $x(t)$ as

$$
x(t) = x_h(t) + x_r(t) \simeq m_h \sin(2\pi f_h t) + m_r \sin(2\pi f_r t) \tag{4.6}
$$

where $x_h(t)$ and $x_r(t)$ represent the body movement caused by the heartbeat and by the respiration, respectively.

These two signals can be approximated by two sinusoidal signals with amplitudes m_h and m_r, respectively, and frequencies f_h and f_r.

If we replace (4.6) in (4.5) we have

$$
\begin{aligned}
s_I(t) &= \cos\left[\frac{4\pi x_h(t)}{\lambda} + \frac{4\pi x_r(t)}{\lambda} + \alpha\right] \\
s_Q(t) &= \sin\left[\frac{4\pi x_h(t)}{\lambda} + \frac{4\pi x_r(t)}{\lambda} + \alpha\right]
\end{aligned}
\tag{4.7}
$$

where $\alpha = \Delta\phi(t) + \theta$ From (4.7) as in [7] we can write

$$\begin{aligned} s_I(t) &= \sum_{p=-\infty}^{+\infty} \sum_{q=-\infty}^{\infty} \cos(2p\pi f_r t + 2q\pi f_h t + \alpha) J_p\left(\frac{4\pi m_r}{\lambda}\right) J_q\left(\frac{4\pi m_h}{\lambda}\right) \\ &= \sum_{p=-\infty}^{+\infty} \sum_{q=-\infty}^{\infty} C_{p,q} \cos(2\pi f_{p,q} t + \alpha) \end{aligned} \tag{4.8}$$

and

$$\begin{aligned} s_Q(t) &= \sum_{p=-\infty}^{+\infty} \sum_{q=-\infty}^{\infty} \sin(2p\pi f_r t + 2q\pi f_h t + \alpha) J_p\left(\frac{4\pi m_r}{\lambda}\right) J_q\left(\frac{4\pi m_h}{\lambda}\right) \\ &= \sum_{p=-\infty}^{+\infty} \sum_{q=-\infty}^{\infty} C_{p,q} \sin(2\pi f_{p,q} t + \alpha) \end{aligned} \tag{4.9}$$

where $C_{p,q} = J_p\left(\frac{4\pi m_r}{\lambda}\right) J_q\left(\frac{4\pi m_h}{\lambda}\right)$ represents the amplitude of the cosinusoinal/sinusoidal component with frequency $f_{p,q} = pf_r + qf_h$ and J_p is the Bessel function of the third kind of order p.

We can combine the two I and Q signals into the complex signal $z(t)$ such that

$$z(t) = s_I(t) + js_Q(t) + w(t) = \sum_{p=-\infty}^{+\infty} \sum_{q=-\infty}^{\infty} C_{p,q} e^{j2\pi f_{p,q} t} e^{j\alpha} + w(t) \tag{4.10}$$

where $w(t)$ represents the noise. It is evident that, since the modulus of the term $e^{j\alpha}$ is always unitary, the problem of the null point detection due to the effect of α does not exist anymore.

This technique is also robust against dc offset, caused by circuit imperfection and the reflections from stationary clutter (objects) surrounding the human body, because it can be estimated in each time window as average of the complex signal and easily removed.

The dc offset, if not cancelled, may sometimes saturate the dynamic range of the baseband filters/amplifiers [8].

But how can the radar estimate the rhythm of physiological movement of the human body from the output signal $z(t)$?

The signal $z(t)$ is sampled and converted into a digital signal $z(n)$ through an ADC device, then processed to get the information on f_r and f_h.

The sampled version $z(n)$ can be written as:

$$
\begin{aligned}
z(t) = z(nT) &= \sum_{p=-\infty}^{+\infty} \sum_{q=-\infty}^{\infty} C_{p,q} e^{j2\pi f_{p,q} nT} e^{j\alpha} + w(nT) \\
&= \sum_{p=-\infty}^{+\infty} \sum_{q=-\infty}^{\infty} B_{p,q} e^{j2\pi f_{p,q} n} + w(n)
\end{aligned}
\tag{4.11}
$$

where T is the sampling interval, $n = 0, \ldots N - 1$, $B_{p,q} = C_{p,q}e^{j\alpha}$, and $f_{p,q} = f_{p,q}T$ is the normalized frequency.

The N-length sequence of samples $z(n)$ theoretically contains an infinite number of complex exponential components whose parameters (frequency and complex amplitude) are unknown. Actually, fortunately, the number of non-negligible components is low, since the power of the harmonics decays very quickly, so we can assume that the sequence $\{z(n)\}_{n=0}^{N-1}$ contains a finite number K of strong components with $K<<N$. From experimental tests [5], it results that K=5 when detecting from the front of the subject, whereas K=4 when detecting from the back. With this approximation the signal $z(n)$ is given by

$$
z(n) = \sum_{k=1}^{K} A_k e^{j2\pi\beta_k n} + v(n) \tag{4.12}
$$

where A_k and β_k are the amplitudes and phases of the strongest components and $v(n)$ is due to the contribution of the noise $w(n)$ and of the weakest components of the signal.

The easiest way to estimate the set of unknown parameters $\{A_k, \beta_k\}_{k=1}^{K}$ from $\{z(n)\}_{n=0}^{N-1}$ is to calculate through the FFT the periodogram of the data.

Ideally, with an infinite length ($N \to \infty$) the spectrum of the signal $z(n)$ would be a sequence of K lines at frequencies β_k, but, as known, the finiteness of N causes smearing and leakage in the calculated periodogram, as shown in Figure 4.2, and frequencies closer than the limit $\Delta f = 1/N$ cannot be resolved since they merge in only one mainlobe. As a matter of fact in Figure 4.2 the two lines around 0.1 Hz are merged in only one lobe, and the weak line at 0.4 Hz is covered by the sidelobes of the strong line at 0.35 Hz.

This means that in some cases the mainlobe of harmonics due to the heartbeat can merge with the mainlobe of components due to the respiration body displacement, or some sidelobes of strong components can cover a weaker mainlobe of weaker components. For example, the third- and fourth-order harmonics of the respiration signal are very close to the fundamental frequency of

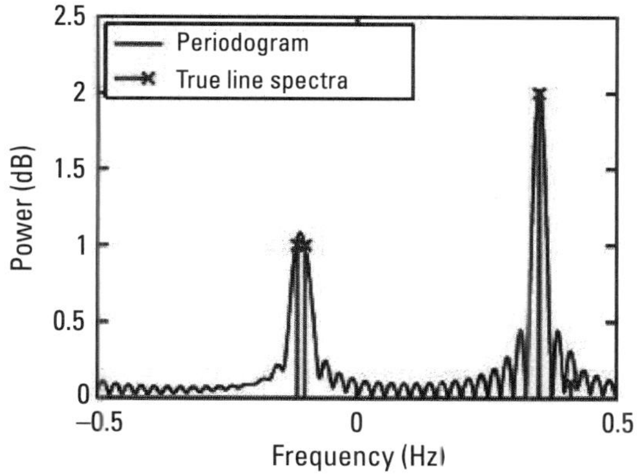

Figure 4.2 Example of periodogram [7].

the heartbeat, complicating the estimation. That's why other high-resolution spectral estimation methods have been proposed for the heartbeat and respiration frequencies estimation: the RELAX algorithm, for instance, whose effectiveness for the problem at hand has been proved in [7]. Some results of this experiment are summarized in Figure 4.3.

The RELAX algorithm in this figure shows that the respiration consists of two closely located sinusoids at 19.1 and 23.5 beats/min due to the variation of respiration rate during the measurement. The heartbeat signal has a frequency of 65.9 beats/min. The other two strong components near 40 beats/min are due to the harmonics of the respiration signal and the intermodulation between respiration and heartbeat signals.

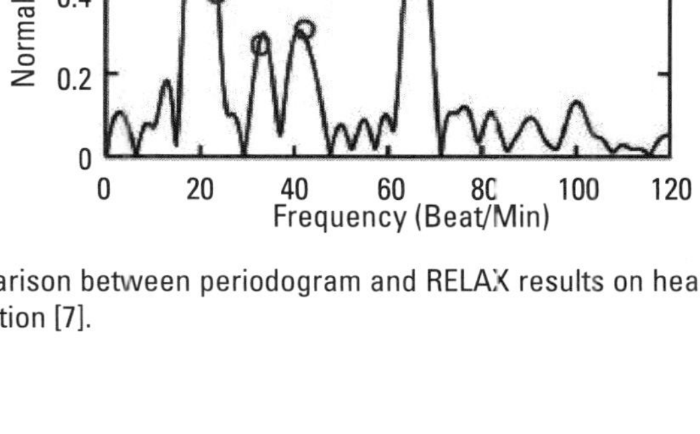

Figure 4.3 Comparison between periodogram and RELAX results on heartbeat and respiration rhythm estimation [7].

4.3 Choice of Carrier Frequency

A general observation on microwave Doppler radar is that the higher the carrier frequency, the shorter the wavelength and the higher the detection sensitivity to small movement. That's why most of the Doppler radar for monitoring vital signs have been designed to work from 450 MHz to 1.8 GHz and 2.5 GHz [8] and recently at 24 GHz [9], 35 GHz [10], and 60 GHz [11].

Unfortunately, the radar detection sensitivity is heavily affected by the nonlinear relation between the received signal $s(t)$ and the useful signal $x(t)$, as shown in (4.8–4.9). As a matter of fact, due to the simultaneous presence of heartbeat and respiration movements, in the complex demodulated signal $y(t)$ many intermodulation components are present as shown in (4.6–4.8). Moreover, even the amplitude of the fundamental term of heartbeat at frequency $f_h = f_{0,1}$ or respiration at frequency $f_r = f_{1,0}$ depends on both indexes m_r and m_h $\left(C_{0,1} = J_0\left(\frac{4\pi m_r}{\lambda}\right)J_1\left(\frac{4\pi m_h}{\lambda}\right) \text{ and } C_{1,0} = J_1\left(\frac{4\pi m_r}{\lambda}\right)J_0\left(\frac{4\pi m_h}{\lambda}\right)\right)$.

Typically for relaxed human beings, m_h is on the order of 0.01 mm and m_r from the order of 0.1 mm to several millimeters. Then, depending on the values of f_h and f_r, m_h and m_r in the analyzed subject, depending on the radar wavelength λ, some of the intermodulation terms can be quite close to the fundamental terms (heartbeat and respiration) that we want to monitor and their amplitude nonnegligible with respect to the useful signal.

In [8], the authors performed a detailed simulated study on the impact of the carrier frequency on the amplitudes of fundamental and intermodulation terms for different values of of f_h and f_r, m_h and m_r. The results show that the best choice for the carrier frequency is not to increase it boundless, but only up to the lower region of Ka-band (between 27 and 30 GHz).

4.4 Phase Noise and Range-Correlation

We have seen in (4.5–4.12) that in the recorded signal a residual phase noise is always present and represents the difference between the phase of the transmitted signal and the received one after demodulation by using a LO signal that is derived by the same source as the transmitted signal. This means that the phase of transmitted and demodulated signals are correlated with the level of correlation dependent on the time delay between the two signals. This correlation effect, known as range-correlation effect, greatly reduces the impact of the phase noise on the received signal. As a matter of fact, if two different oscillators with uncorrelated phase noise were used for transmitting and receiving, it would be impossible to detect the small phase variations created by heart and respiration motion.

Range-correlation theory describes how the residual phase noise affects the baseband noise spectrum. According to [4], with the target at a given range, the baseband noise spectral density is given by

$$S_{\Delta\phi}(f) = 4S_{\phi}(f)\sin^2\left(\frac{2\pi Rf}{c}\right) \tag{4.13}$$

where $S_{\phi}(f)$ is the RF phase-noise power spectral density and the sinusoidal term is due to the range-correlation effect.

At values relevant for radar monitoring of heartbeat and respiration, Rf/c will be very small, on the order of 10^{-9}, so the small-angle approximation is valid and the baseband noise spectrum will increase proportionally to the square of the target range and the square of the offset frequency as shown in (4.14).

$$S_{\Delta\phi}(f) = 4S_{\phi}(f)\left(\frac{2\pi Rf}{c}\right)^2 \tag{4.14}$$

The effect of this phase-noise has been analyzed in [4], showing that, for instance, at a 50-cm range and a frequency of 10 Hz, it is 134 dB below the RF phase noise power.

Range correlation effect is then particularly important in measuring the chest-wall movement, since it is encoded in a signal $x(t)$ with frequencies in the range of 0.1–10 Hz where the RF phase noise is very high.

More considerations on noise sources and limitations in CW radar for vital signs monitoring are presented in [12].

4.5 Front-End Architectures

The front-end architecture plays an important role in maximizing the sensitivity, extending the detection range, and rejecting interferences. Various RF front-end architectures have been proposed over the years. Historically, the first architecture for noncontact vital sign monitoring has been the heterodyne, and it was the only one until the homodyne detector was proved effective at the beginning of last decade. Other very recent architectures are the double-sideband, direct IF sampling, and self-injection locking. The properties of some of these front-end architectures are discussed in this section.

4.5.1 Homodyne

A simplified block diagram of homodyne detector is illustrated in Figure 4.4. There are the two I and Q baseband outputs to solve the problem of the null detection point. The two output narrowband channels are then combined to perform complex signal processing, as shown in Section 4.2, or arctg (AT) demodulation [4].

4.5.2 Double-Sideband Heterodyne

Compared with the homodyne architecture, the heterodyne architecture has the advantage of robustness against dc offset. However, single-channel heterodyne radar has the null detection point problem [13]. In order to overcome the null detection point problem, double sideband heterodyne architecture is generally used. The block diagram of the double-sideband heterodyne detector is shown in Figure 4.5.

As shown in Figure 4.5, the upper sideband and the lower sideband frequency tones are generated and transmitted by mixing signals from two voltage-controlled oscillators at RF (LO1) and IF (LO2). At the receiver, the two sidebands are combined automatically by two-stage down-converters. The signal detected by either the upper sideband or the lower sideband has null detection and optimal detection points separated by $\lambda/8$. When the two sidebands are combined, the distance between optimal and null detection points becomes $\lambda_{IF}/16$, where λ_{IF} is the wavelength corresponding to the IF, thus resulting in a much longer separation. Moreover, the location of optimal/null detection points can be changed based on the IF [13], and the double-sideband transceiver architecture can always avoid the null detection point problem by IF tuning. This is at the same time an advantage and a disadvantage of the

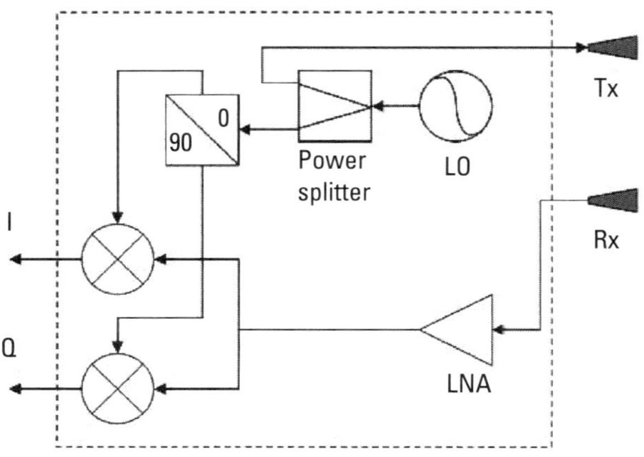

Figure 4.4 Block diagram of a homodyne detector.

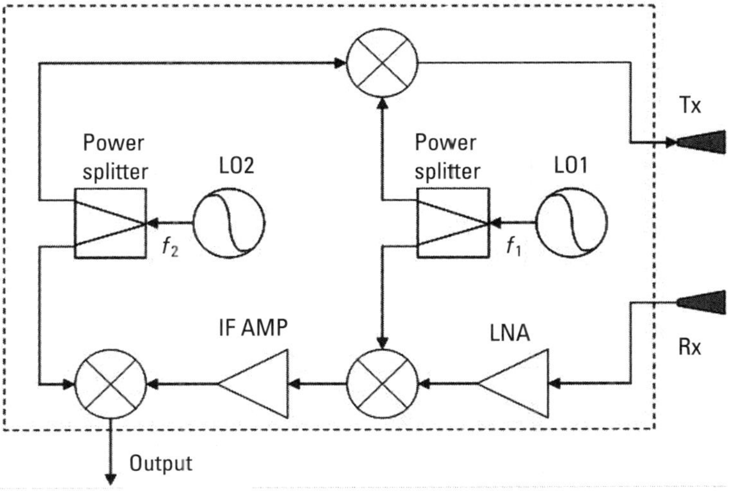

Figure 4.5 Double-sideband detector.

double-sideband architecture, since each time the detection distance changes, a sophisticated procedure to operate the system or automatic calibration function is required, increasing the cost of the device.

The double-sideband architecture provides an advantage when it comes to monolithic integration. Without the need for image rejection filters and quadrature generation, the radar transceiver can be easily designed and integrated on a semiconductor chip.

4.6 UWB Radar for Health Care Monitoring

Applying UWB technology in medical applications is an emerging research trend in recent years.

The first attempt at using UWB radar in medical applications was in 1993 for human body monitoring and imaging. In 1996, the biomedical use of UWB radars was better described and in the same year, the first US patent was awarded. By 2000, many projects around the word focused on UWB medical applications in cardiology, obstetrics, breath pathways, and arteries. This section of the chapter is dedicated to the applications of UWB radars for the monitoring of heartbeat and respiration rhythm.

In the ultrawideband radars, the transmitted signal bandwidth exceeds "the lesser of 500 MHz or 20% of the center frequency," according to the Federal Communication Commission (FCC) definition [14]. This transmit wideband can be obtained in different ways, either using nonsinusoidal pulses of a typical width of less than 1.5 ns, and thus a bandwidth over 1 GHz, or using longer, phase-coded pulses and pulse compression. Among pulses with definite mathematical expressions, typically used pulses include Gaussian monocycles, Gaussian doublet, Rayleigh monocycles, and Manchester monocycles. A general discussion on UWB pulses can be found in [15] and in Chapter 5.

Currently, the only available emission masks for UWB radio are those issued by the FCC in the United States [14] (as shown in Figure 4.6). The FCC's regulation established three types of UWB devices based on their potential to cause interference: (1) imaging systems including ground penetrating radars, wall-through-wall surveillance, and medical imaging devices; (2) vehicular radar systems; and (3) communications and measurement systems.

The regulation mentions that medical systems must be operated in the frequency band 3.1–10.6 GHz. and describes it as a system that "may be used for a variety of health applications to 'see' inside the body of a person or animal." And, it specifically says, "Operation must be at the direction of, or under the supervision of, a licensed care practitioner."

In a UWB pulse radar, the transmitter sends very short electromagnetic pulses toward the target. The typical pulse duration for vital sign detection is

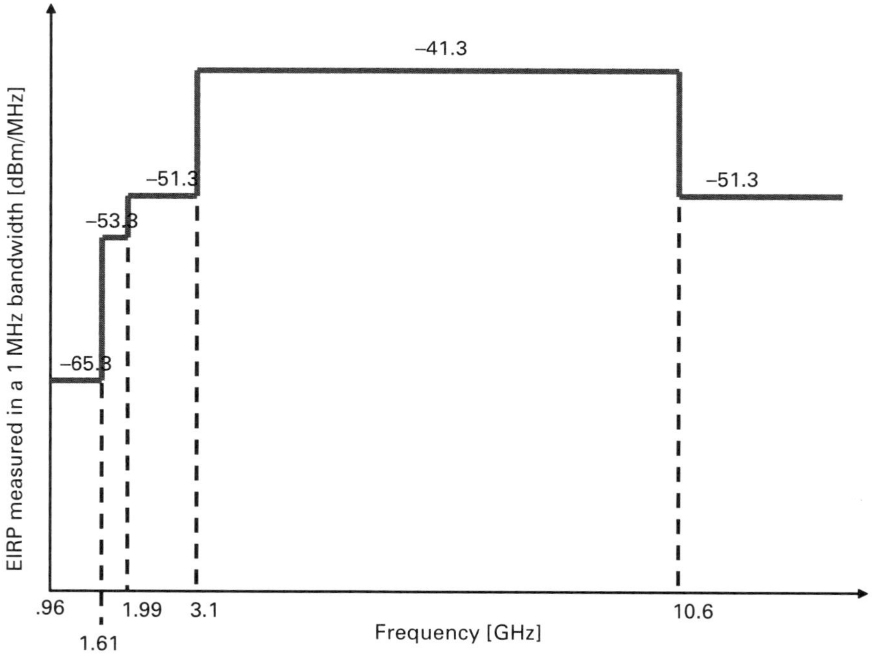

Figure 4.6 UWB mask. Emission limits for ground penetrating radar, wall imaging and medical imaging systems [14].

around 200–300 ps, and the pulse repetition frequency is in the range of 1–10 MHz.

When the transmitted pulse reaches the chest wall, part of the energy is reflected and captured by the receiver. If a local replica of the transmitted pulse with a delay close to the nominal round-trip traveling time $t_0 = 2R_0/c$ is correlated with the received echo, the output correlation function will have the same frequency as the physiological movement.

By controlling the delay between the two inputs of the correlation function block, the detection range of UWB radar can be changed since the delay corresponds to the signal round-trip traveling time. This mechanism makes it possible for UWB radar to eliminate interference caused by reflection from other objects (clutter) and multipath reflection. This is actually the main advantage of UWB vital sign detection radar. However, the disadvantage of UWB radar is that the delay needs to be calibrated each time the detection distance changes, with an increased system complexity and cost.

4.7 UWB Radar with Correlator

Figure 4.7 shows the block diagram of the UWB pulse radar sensor [6] with a correlator. The pulse generator transmits short pulses toward the human body with a pulse repetition frequency (PRF) fR. The signals reflected by the target are captured by the RX antenna, amplified by the LNA and correlated (i.e., multiplied and integrated) with a delayed replica of the transmitted pulses generated on-chip by the shaper. The signal at the output of the integrator is captured to recover the information on the movement rate. Since vital signs vary within a few Hertz, an integrator with a 3-dB band (B_{3dB}) of 100 Hz can allow an accurate detection. In practice, the integrator (Int, in Figure 4.7) is implemented by means of a low-pass filter with bandwidth B3dB.

The operating principle can be explained intuitively as follows. For simplicity, let us assume that the delay generator provides a delay equal to the entire round trip time of the transmitted and received pulses. If the target is not

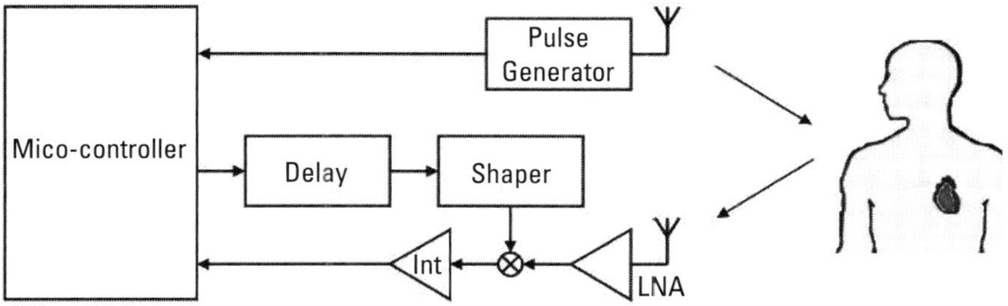

Figure 4.7 Block diagram of the UWB pulse radar sensor.

static, the local replica and the amplified echo are aligned, and the multiplier provides the same output pulse with pulse repetition frequency f_R, as shown in Figure 4.8(a) where $v_0(t)$ is the multiplier output and $v_{out}(t)$ is the integrator output. Therefore, the signal at the output of the integrator is almost constant. It is worth noting that, if the target is still, the integrator will provide an almost constant output voltage (positive, negative, or even zero), regardless of the relative shift between the local replica and amplified echo [16] since the result of the product between received signal and generated one by the shaper is the same. Conversely, if the target is moving, the movement causes a time-varying delay between the local replica and the eco amplified by the LNA. The signal $v_0(t)$ is not always the same since its shape depends on the relative shift between received and shaper signal. Therefore, the output of the integrator $v_{out}(t)$ is no longer constant, and its variations are related to the frequency of heartbeat and respiration—that is, to the signal $x(t)$ as in (4.6). As a matter of fact, if the position of the maximum of the correlation function for each pulse corresponds to the actual target distance, then the output amplitude of the integrator is proportional to the signal $x(t)$ [17], and then the frequencies of heartbeat and breath can be estimated through the Fourier transform of the signal at the output of the integrator. Before any processing, the signal $v_0(t)$ is sampled and converted into a digital signal $v_{out}(n)$ through an ADC device and then processed to get the information on f_r and f_h.

An example of this signal is plotted in Figure 4.9 taken from [17], where the breathing of a person with a breath hold of 12s at 35s can be clearly seen. The heartbeat ripples are evident particularly during the time in which the breath is held.

A different method to measure movements would be to use the UWB radar in tracking mode where the target can be, for instance, the heart. Since the transmitted radar signal penetrates into the human body and the boundary between the heart muscle, and since the surrounding fat tissue has a significantly different permittivity, the heart could be detected and its movements monitored. Currently, however the monitoring of the heart muscle seems to be still

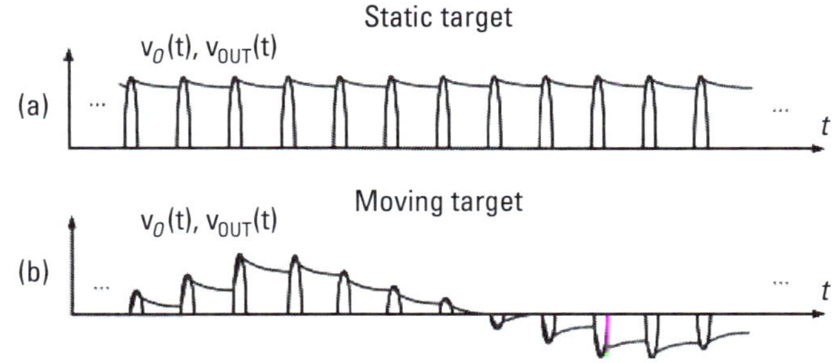

Figure 4.8 Output of UWB pulse radar sensor [16].

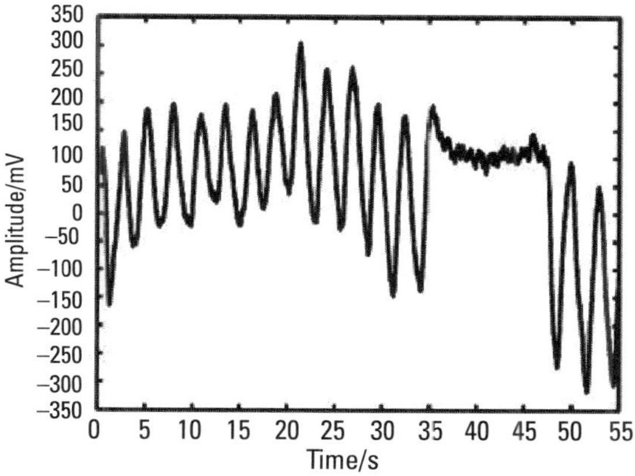

Figure 4.9 Measured sensor output for a breathing pe-son and a breath hold of about 12s [17].

unfeasible, as the signal is strongly attenuated and therefore disappears within the ringing signal of the main target, the skin-air boundary.

A model for the signal power attenuation through the body tissues has been proposed by Staderini in [18] and is summarized in Figure 4.10 This model shows, for instance, a 20-dB round trip loss for the signal to reach the heart and come back with a continuous wave at 1.5 GHz.

4.8 Conclusions

The aim of this chapter was to introduce the main statistical signal processing techniques used in vital signs monitoring performed by both CW and UWB radars. The working principles of both systems is very easy, but still several challenges need to be solved.

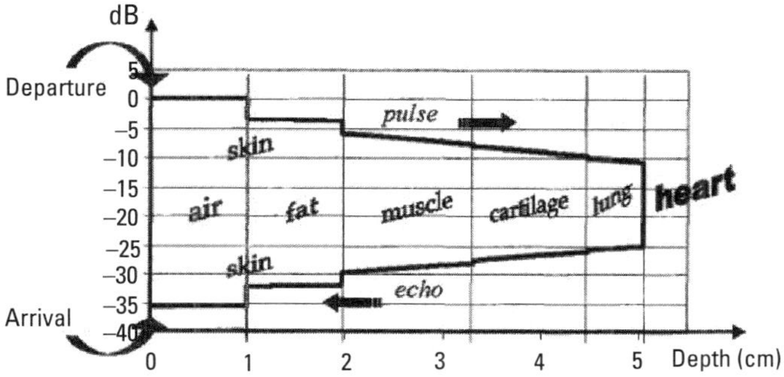

Figure 4.10 Model predicted attenuation of pulse-echo intensity traveling from the transmitting antenna to the receiving antenna [18].

One significant challenge is, for instance, the random body motion artifact cancellation. While the radar sensor is sensitive enough to pick up tiny physiological movements due to breathing and heartbeat, it can also pick up any physical body movement. The use of multiple sensors or MIMO radars has been proposed to mitigate the problem [19].

Another issue can be the presence of scattering due to steady objects that translates to a strong DC component that can saturate the radar detector.

In the UWB radars based on correlator device, conventional coherent detection may sometimes be unsuitable since the UWB pulses traveling through the tissues are deformed. Then, with these signals, unless adaptive template signals could be constructed and updated in real time, the correlation procedure does not provide the expected results.

Anyway, although the technology is still in the development phase, the nonintrusivity, high accuracy, and attractive features of both CW and UWB radar noncontact vital sign detection indicate that it will soon become a part of daily life for many people.

References

[1] Lin, J. C., "Noninvasive Microwave Measurement of Respiration," *Proc IEEE*, Vol. 63, No. 10, Oct. 1975, p. 1530.

[2] Lin, J. C., "Microwave Sensing of Physiological Movement and Volume Change: A Review," *Bioelectromagnetics*, Vol. 13, No. 6, 1992, pp.557–565.

[3] Chen, K. M., D. Misra, H. Wang, H. R. Chuang, and E. Postow, "An X-Band Microwave Life-Detection System," *IEEE Trans. Biomed.Eng.*, Vol. 33, July 1986, pp. 697–702.

[4] Droitcour, A. D., O. Boric-Lubecke, V. M. Lubecke, and J. Lin, et al., "Range Correlation and I/Q Performance Benefits in Single-Chip Silicon Doppler Radars for Noncontact Cardiopulmonary Monitoring," *IEEE Trans. Microw. Theory Tech.*, Vol. 52, No. 3, Mar. 2004, pp. 838–848.

[6] Fouladi, R. F., and A. Oncu, "Vital Signs Modeling for Doppler Radar Cardiorespiratoty Monitoring," *2013 36th International Conference on Telecommunications and Signal Processing*, 2013, pp. 363–366.

[7] Li, C., J. Ling, J. Li, and J. Lin, "Accurate Doppler Radar Noncontact Vital Sign Detection Using the RELAX Algorithm," *IEEE Trans. on Instrumentations and Measurement*, Vol. 59, No. 3, March 2010, pp. 687–695.

[8] Li, C., and J. Lin, "Optimal Carrier Frequency of Non-Contact Vital Sign Detectors," *Proc. IEEE Radio Wireless Symp.*, 2007, pp. 281–284.

[9] Vinci, G., S. Lindner, F. Barbon, and S. Mann, et al., "Six-Port Radar Sensor for Remote Respiration Rate and Heartbeat Vital-Sign Monitoring," *IEEE Trans. on Microwave Theory and Techniques*, Vol. 61, No. 5, Pt. 2, 2013, pp. 2093–2100.

[10] Chioukh, L., H. Boutayeb, L. Lin, and L. Yahia, et al., "Integrated Radar Systems for Precision Monitoring of Heartbeat and Respiratory Status," *Proc. of 2009 Asia Pacific Microwave Conference*, 2009, pp. 405–408.

[11] Kao, T.-Y. J., and J. Lin, "Vital Sign Detection Using 60-GHz Doppler Radar System," *Proc. of 2013 IEEE International Wireless Symposium*, 2013.

[12] Jensen, B. S., T. Jensen, V. Zhurbenko, and T. K. Johansen, "Noise Considerations for Vital Signs CW Radar Sensors," *Proc. of 5th European Conference on Antennas and Propagation*, 2011, pp. 2805–2809.

[13] Jensen, B. S., T. K. Johansen, and L. Yan, "An Experimental Vital Signs Detection Radar Using Low-IF Heterodyne Architecture and Single-Sideband Transmission," *Proc. of 2013 IEEE International Wireless Symposium*, 2013.

[14] Federal Communications Commission, "Revision of Part 15 of the Commission's Rules Regarding Ultra-Wideband Transmission Systems." ET Docket 98-153, Washington, DC, April 22, 2002.

[15] Kissik, W. (ed.), *The Temporal and Spectral Characteristics of Ultrawideband Signals.* NTIA Report 01-383, US Department of Commerce, 2001.

[16] Zito, D., D. Pepe, M. Mincica, and F. Zito, "A 90nm CMOS SoC UWB Pulse Radar for Respiratory Rate Monitoring," *2011 IEEE Solid-State Circuits Conference Digest of Technical Papers*, 2011, pp. 40–41.

[17] Leib, M., W. Menzel, B. Schleicher, and H. Schumacher, "Vital Signs Monitoring with a UWB Radar Based on a Correlation Receiver," *2010 Proceedings of the 4th European Conference on Antennas and Propagation*, 2010, pp. 1–5.

[18] Staderini, E.M., "UWB Radars in Medicine," *IEEE Aerospace and Electronic Systems Magazine*, Vol. 17, No. 1, Jan. 2002, pp. 13–18.

[19] Li, C., and J. Lin, "Random Body Movement Cancellation in Doppler Radar Vital Sign Detection," *IEEE Trans. Microwave Theory Tech.*, Vol. 56, No. 12, Dec. 2008, pp. 3143–3152.

5

Radar for Automotive Applications: Signal Processing Perspective

5.1 General Characteristic of the Sensor and Its Functions

Radar is a key technology for automotive driver assistance applications due to its inherent advantages like weather independence and direct acquisition of range and velocity, especially when compared to alternative sensors like video, laser, and ultrasonic. In addition radar offers the advantage of being invisibly mounted behind electromagnetic transparent materials (e.g., plastic bumpers).

In 1999 Mercedes Benz was the first car manufacturer to introduce a radar-based autonomous cruise control (ACC) system in its S-class. Nowadays, the global auto industries are extensively pursuing radar-based proximity detection systems for applications including ACC, collision avoidance (CA), precrash warning, mitigation of collision damage, and "stop and go" functionalities and parking aid (forward and reverse) with rear-crash collision warning, as shown in Figure 5.1.

Then, automotive radar systems need to have the capability to measure range, velocity, and azimuth angle simultaneously for all points and extended targets inside the observation area with short measurement time (for ACC about 100 ms, for CA much shorter), even in dense target situations, with high range accuracy and resolution in all environments, ranging from scorching heat to snowstorms. They must be small, lightweight, and low cost.

There are many constraints for an automotive radar system. Styling is a critical parameter and a radar sensor, even if very efficient, will be unacceptable if it cannot be styled into the vehicle package. The front and back grille area

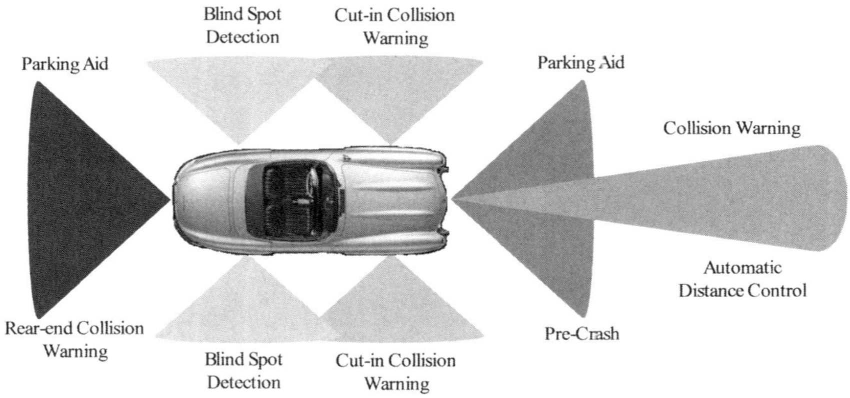

Figure 5.1 Automotive radar functionalities.

has been the traditional placement area, but also the area around the rear view mirror is used.

The choice of the antenna has a significant impact on packaging, cost, and performance of the sensor. A wide range of solutions has been used over the past 15 years: single narrow-beam antennas, horn antennas, static planar antennas, mechanically and electronically scanned planar arrays, twist polarization Cassegrain reflectors, staring beam antennas, and switched beam antennas [1]. Mechanical scanning systems offer the benefit of optimizing the beam shape, which does not change as the beam is scanned. Electronic scanning offers the possibility of moving the antenna beam without physical movement; however, since the apparent aperture width changes as the beam is scanned, the beam performance also changes.

The choice of operating frequency is dictated by national and international bodies. In radar engineering terms, the choice of operating frequency is linked with the antenna beamwidth and the antenna size [1]: the higher the frequency, the smaller the antenna for the same beamwidth. This logic made the use of high frequencies very attractive. As a matter of fact, the modern automotive radar works at 24 and 76–77 GHz. But, although it seems futuristic today, the area between 152–154 GHz or even around 228 GHz is under consideration for future automotive applications, resulting in either dramatic package shrinking or largely enhanced angular resolution using either digital beam forming or holographic principles [2].

Based on the maximum distance covered by the radar, the automotive radar systems are generally divided in long-range radar (LRR) and short-range radar (SRR), as shown in Figure 5.2.

At present in more than 70 countries the frequency 76–77 GHz is allocated for LRR. Most of them are frequency linearly modulated continuous

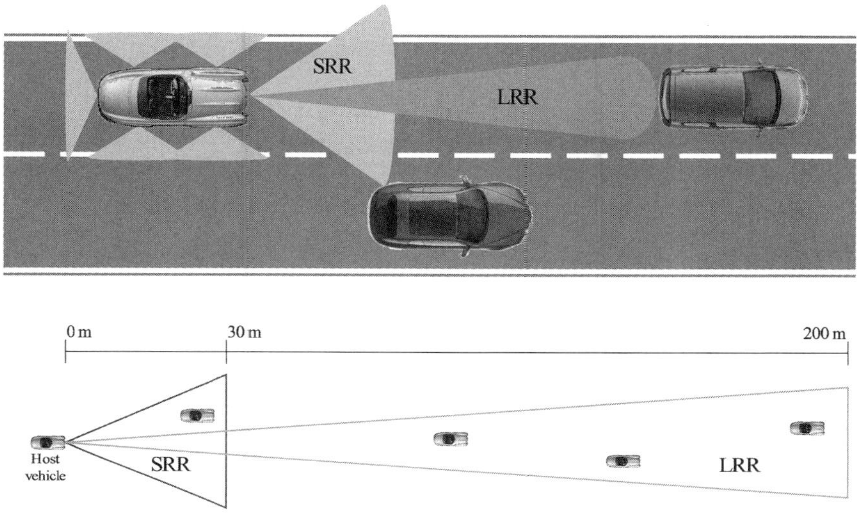

Figure 5.2 SRR and MRR radars.

waveform (FLMCW) radar, but there are also other solutions using classical pulse waveform with ultra short pulse length (10 ns) [3]. In addition, there is continuing research on using spread spectrum transceivers that have a combined use of intervehicle communications as well as radar ranging [4].

The main functionality of a LRR is the ACC. It must detect targets of varying size (small children, trucks, and so on), and it covers a maximum distance of 80–200 m with an angle of 8–10 degrees. This range and azimuth coverage is needed to track the targets in the lanes ahead in order to anticipate vehicles cutting in front of the car and in order to distinguish between targets. For example, when driving toward a curve, the ACC system must distinguish between the target vehicle, sign posts, guard rails, and stopped vehicles [3]. The typical specifications of ACC radar sensor are summarized in Table 5.1.

The SRR covers a maximum distance of 30m, with an angle of about 70 degrees and is used for parking assistance, precrashing sensing, and stop and go. The frequency of 24 GHz is allocated for SRR sensors and, because of the necessity to achieve high local resolution to distinguish between objects that are close to each other with an accuracy of centimeters, they often use the UWB technology.

The general characteristics of LRR and SRR are summarized in Table 5.2.

5.2 Signal Processing for the Single Sensor

The main blocks of a LRR radar are summarized in Figure 5.3.

Table 5.1
Typical Performance Specification for ACC Radar Sensor

Transmit frequency	76–77 GHz
Transmit power	>10 dBm
Target detection distance (m)	
Range	2-150
Accuracy	< 1 or
Relative velocity (Km/h)	
Range	-250 +250
Accuracy	> 1
Angular coverage (°)	
Azimuth	-8 +8
Elevation	3–4 single beam
Update rate	> 10Hz

Table 5.2
Automotive Radar Specifications

	Modul.	Freq. (GHz)	Bandwidth (GHz)	Max Range (m)	Resol. (m)	View Angle (Degrees)
SRR	Pulsed	24	7	30	<0 1	70
SRR	Pulsed	79	4	30	<0 1	70
LRR	FMCW	76-77	1	200	1	10

A 77-GHz voltage controlled oscillator (VCO) generates the required transmitted FMCW signals to the transmit antenna. Part of the transmitted signal is coupled back to the LO input of a millimeter wave mixer. The echoes from potential targets are received by the receiving antenna and inputted into the millimeter wave mixer. The mixer produces both the in-phase (I) and quadrature-phase (Q) IF components. These signals are then digitized by the digitizer (ADC) and inputted into DSP board. The DSP board also controls the operation of the VCO by feeding the DAC circuit commands to produce the modulating voltage.

The design of the modulation waveform is very important, as it is one of the factors that determine the performance of the radar. A detailed analysis of many waveforms is performed in [5]. In this chapter, we report the results for only the most common one, the triangular chirp.

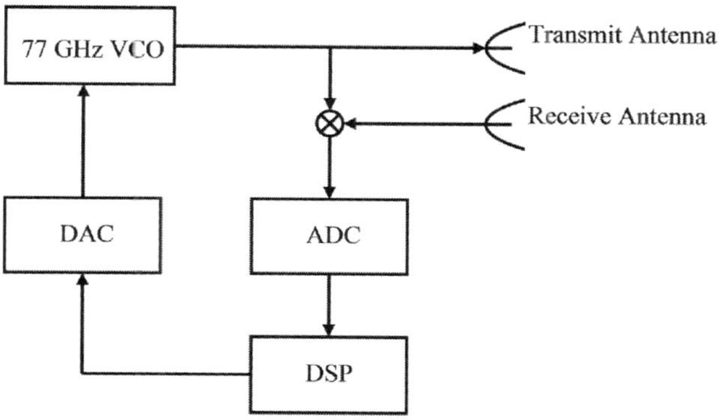

Figure 5.3 Block diagram of LRR.

5.2.1 Range and Frequency Estimation

In an FMCW radar using triangular modulation, the continuously transmitted waveform is a modulated sinusoidal signal

$$s_T(t) = \cos\left[2\pi\left(f_T t \pm \frac{B_{sw}t^2}{2T_{CPI}}\right)\right] \tag{5.1}$$

where f_T is the carrier frequency, B_{sw} is the sweep bandwidth, and T_{CPI} is the chirp time duration. The positive slope of the signal indicates an up-chirp signal and the instantaneous frequency is given by $f_u(t) = f_T + \frac{B_{sw}t}{T_{CPI}}$ while the negative slope a down-chirp signal with the same bandwidth Bsw, and the instantaneous frequency is given by $f_d(t) = f_T - \frac{B_{sw}t}{T_{CPI}}$. In the first TCPI the up-chirp is transmitted, followed in the second TCPI by a down-chirp, as shown in Figure 5.4.

If in the illuminated area a target is present, moving with a radial velocity v_r constant during the time-on-target (ToT), the time-dependent range $R(t)$ changes such that

$$R(t) = R_0 + v_r t \tag{5.2}$$

and the signal propagation delay is

$$\tau(t) = \frac{2}{c}R(t) = \frac{2}{c}(R_0 + v_r t) \tag{5.3}$$

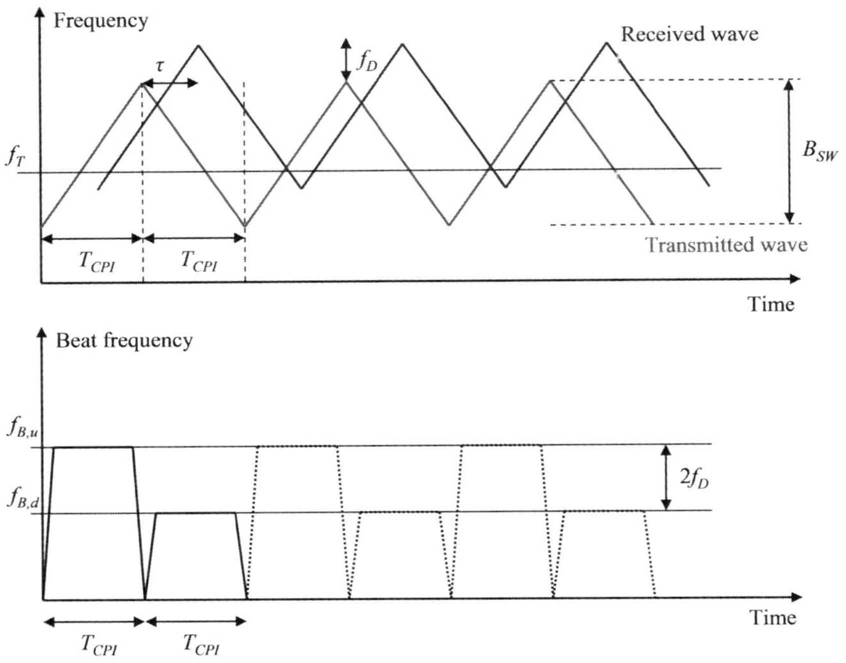

Figure 5.4 Triangular chirp.

where c is the light speed (3 · 108 m/s), and R_0 the starting position of the target. Then, the time-delayed received target signal $s_R(t)$ is proportional to $s_T(t - \tau(t))$ and is directly down-converted into the baseband by the instantaneous transmit frequency $f(t)$. The frequency difference between the receive signal $s_R(t)$ and the transmit signal $s_T(t)$ is denoted as beat frequency f_B.

In the case of triangular chirp, two beat frequencies will be measured in the up- and down-chirp complex valued baseband signals, as shown in Figure 5.4.

$$\begin{cases} f_{B,u} = -\dfrac{2}{\lambda}v_r - \dfrac{2B_{sw}}{cT_{CPI}}R = f_D + f_R \\ f_{B,d} = -\dfrac{2}{\lambda}v_r + \dfrac{2B_{sw}}{cT_{CPI}}R = f_D - f_R \end{cases} \tag{5.4}$$

where we used the relation $f_T = c/\lambda$, and we supposed that during the up- and down-CPI the target is still, then $R(t) = R$; f_D is the target Doppler frequency, due to its velocity, and f_R is the contribution from the target range. Solving these two equations the target range R and the velocity v_r can be calculated in an unambiguous way as follows:

$$\begin{cases} R = \dfrac{cT_{CPI}}{4B_{sw}} \dfrac{\left(f_{B,d} - f_{B,u}\right)}{2} \\ v_r = -\dfrac{\lambda}{2} \dfrac{\left(f_{B,u} + f_{B,d}\right)}{2} \end{cases} \tag{5.5}$$

For any beat frequency measurement, the beat frequency resolution is inversely proportional to the CPI; that is, $\Delta f_B = \dfrac{1}{T_{CPI}}$. Conversely, the range resolution is defined as

$$\Delta R = \frac{c}{2}\Delta\tau = \frac{c}{2}\frac{\Delta f_B T_{CPI}}{B_{sw}} = \frac{c}{2B_{sw}} \tag{5.6}$$

Then, the sweep bandwidth B_{sw} is determined by the desired range resolution ΔR, and the T_{CPI} by the frequency resolution Δf_B. A typical range resolution of 1m implies a sweep bandwidth of 150 MHz, and a velocity resolution of 1km/h at 77 GHz implies a frequency resolution of 71 Hz and $T_{CPI} \approx 14$ *ms.*

It is clear that the solution in (5.5) is unambiguous when only a single target is present. In a multiple targets scenario, the triangular chirp provides an up- and down-beat frequency equation for each target; then for two targets there are four different ambiguous solutions. In this case, alternative waveforms have been proposed as explained in [5].

5.2.1.1 How to Measure $f_{B,u}$ and $f_{B,d}$

After the down-conversion in baseband, the received signal is passed through a low pass filter (LPF) to remove the noise and then sampled by the A/D converter.

The resulting discrete time signal, that we denote with $s_D(n)$, $n = 0, 1 \ldots, N-1$, is weighted with windows (Hamming [6] and Blackman [7]) to limit the frequency content, zero-padded, and then an FFT is applied to it for each CPI.

The temporal signal is therefore transformed in to a frequency signal as

$$\begin{aligned} S(k) &= \sum_{n=0}^{N-1} w(n) s_D(n) \exp\left(-j\frac{2\pi nk}{N}\right) \\ &= \sum_{n=0}^{N-1} w(n) s_D(n) W_N^{nk} \text{ with } k = 0,1,\ldots,N-1 \end{aligned} \tag{5.7}$$

where are the window samples and $W_N^{nk} = \exp\left(-j\frac{2\pi nk}{N}\right)$

If there was no disturbance or interference, the frequency signal $S(k)$ would present a neat peak corresponding to the beat frequency for each CPI (up- and down-beat frequency) from which it would be possible to derive R and v_r. Unfortunately, the disturbance is unavoidable, and some peak detection/estimation algorithm must be applied.

An example of the absolute value of $S(k)$ is reported in Figure 5.5 (in arbitrary scale) for the up ($|S_U|$) and down ($|S_D|$) chirp. The two beat frequencies are evident.

5.2.1.2 Some Remarks on the Fast Fourier Transform

The FFT algorithms are based on the fundamental principle of decomposing the computation of the discrete Fourier transform (DFT) of a sequence of length N into successive smaller DFTs. There are many algorithms to implement it, the most popular of which are the decimation-in-time and the decimation-in-frequency [8].

The flow graph of the decimation-in-time decomposition of an eight-point DFT is plotted in Figure 5.6 with the typical butterfly computation. In this figure, $x(n)$ represent the time samples, and $X(k)$ the DFT frequency samples.

The coefficients W_N^{nk} are the key elements of the FFT computation, since they are symmetric and periodic. These properties allow the re-use of the coefficients in different stages of the algorithm, as shown in Figure 5.6. The number of complex multiplications and additions for this FFT is of the order of $N\log_2 N$, while for the classical DFT algorithm they are N^2.

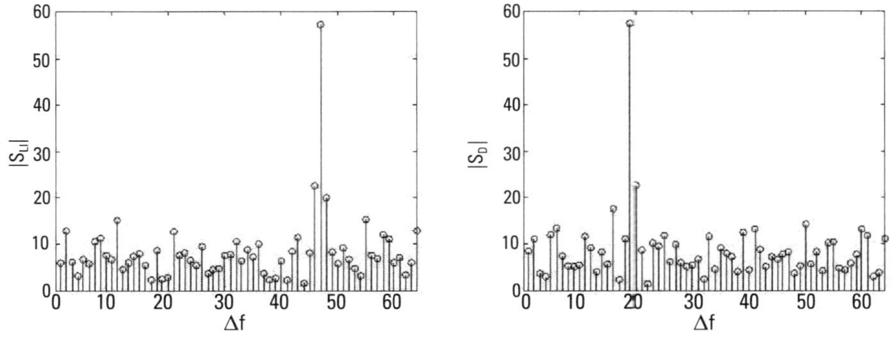

Figure 5.5 Plot of the absolute value of S(k) for up- and down-chirp.

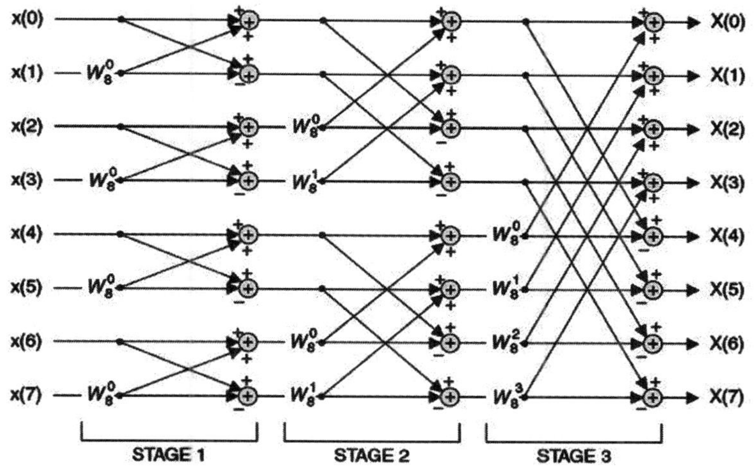

Figure 5.6 Flow graph of the decimation-in-time decomposition of an eight-point DFT.

5.2.2 CFAR Processing

The algorithm that is often applied to detect and estimate the peak in the Fourier transformed signal S(k) is the cell averaging constant false alarm rate (CA-CFAR) algorithm. Other alternatives are the greatest of (GO)-CFAR, the smallest of (SO)-CFAR, the ordered statistic (OS)-CFAR, and the trimmed mean (TM)-CFAR. The general scheme of these CFAR algorithms is shown in Figure 5.7.

This scheme can be applied both on the range and frequency samples of the radar received signal. In the radar at hand, it is applied on the frequency samples to detect the presence of the peak and to estimate its position.

The samples $S(k)$ are first squared providing $X(k) = |S(k)|^2$ and then processed.

Since the power of the disturbance frequency samples is unknown, the fixed-threshold Neyman-Pearson type detector is not applicable here. A common approach is to employ a detector that sets thresholds based on local estimates of the total disturbance power ([9] and references therein). As shown in Figure 5.7, the threshold is computed on a sample-by-sample basis forming a finite reference window of size K surrounding the CUT and using the K reference cells to estimate the unknown power. A certain number N_G of guard cells is avoided in the power calculation in order to reduce target masking.

Then, the square amplitude of the cell under test (CUT) echo Y is compared with a threshold, such that

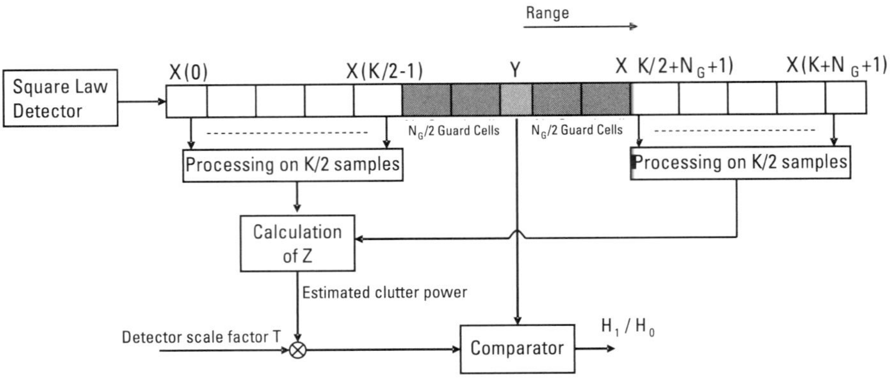

Figure 5.7 General scheme for a CFAR detector.

$$Y \underset{H_0}{\overset{H_1}{\gtrless}} T\,Z \tag{5.8}$$

where T is a deterministic constant scale factor that depends on the desired constant probability of false alarm (P_{FA}) and Z is a threshold that depends on the data surrounding the CUT.

In the case of the CA-CFAR processor with $N_G = 0$, the threshold Z is given by:

$$Z = \frac{1}{K}\left(\sum_{k=0}^{K/2-1} X(k) + \sum_{k=K/2+1}^{K} X(k)\right) \tag{5.9}$$

that is, by the mean level estimate in the K reference cells (or power estimate of the sample $S(k)$ in the reference window). A target (the peak of the spectrum) is declared to be present if Y exceeds the threshold TZ.

The CA-CFAR processor is the optimum CFAR processor [9], since it maximizes the detection probability (P_D) in a homogeneous background where the samples $X(k)$ are independent and identically distributed (i.i.d.) with an exponential distribution. Under these hypotheses, the probability of false alarm is given by

$$P_{FA} = (1 - T)^{-K} \tag{5.10}$$

and, as the size of reference cells K increases, the P_D approaches that of the optimum detector with a fixed threshold. Unfortunately these hypotheses are not always true. For instance, there can be an abrupt change of noise power in the reference cells, or there can be some outliers. The presence of such phenomena may result in severe performance degradation of the CA-CFAR, leading to an excessive false alarm rate or target masking [9, 10].

Modifications of the CA-CFAR detector have been proposed to overcome the problems of nonhomogeneous background. In [11], the author proposed a CFAR procedure to regulate false alarms in the region of clutter transition. In this processor, the noise power is estimated by the greatest of the sum in the leading and lagging windows as:

$$Z = \frac{2}{K}\max\left(\sum_{k=0}^{K/2-1} X(k), \sum_{k=K/2+1}^{K} X(k)\right) \tag{5.11}$$

leading to the so called GO-CFAR. This detector exhibits a loss with respect to the CA-CFAR of 0.1 up to 0.3 dB in homogeneous exponential background. The probability of false alarm in this case, under the same hypothesis of i.i.d exponential samples, is given by

$$P_{FA} = 2(1+T)^{-K/2} - 2\sum_{k=0}^{K/2-1}\binom{K/2+k-1}{k}(2+T)^{-(K/2+k)} \tag{5.12}$$

To prevent the suppression of closely spaced targets (peaks in the frequency domain), Trunk [12] proposed a CFAR detector in which the threshold is calculated as the smallest of the sum in the leading and lagging windows

$$Z = \frac{2}{K}\min\left(\sum_{k=0}^{K/2-1} X(k), \sum_{k=K/2+1}^{K} X(k)\right) \tag{5.13}$$

leading to the so called SO-CFAR. This detector performs very well in resolving two closely spaced targets but in some conditions exhibits great losses in homogeneous background, up to 11 dB for K=4 and PFA=10-6.

The probability of false alarm in this case, under the same hypothesis of i.i.d exponential samples, is given by

$$P_{FA} = 2(1+T)^{-K/2}\sum_{k=0}^{K/2-1}\binom{K/2+k-1}{k}(2+T)^{-k} \tag{5.14}$$

A more complex alternative, called OS-CFAR, has been proposed in [13]. The OS-CFAR detector estimates the noise power simply selecting the *k*th larger sample in the reference window, such that

$$Z = X_{(k)} \text{ where } X_{(1)} \leq X_{(2)} \leq \cdots \leq X_{(K)} \tag{5.15}$$

The probability of false alarm in this case, under the same hypothesis of i.i.d exponentiallly distributed samples, is given by

$$P_{FA} = k\binom{K}{k}\frac{(k-1)!(T+K-k)!}{(T+K)!} \tag{5.16}$$

where k is the order of the selected sample. The OS-CFAR suffers only minor degradation of the PD in homogeneous noise and resolves closely spaced targets [9].

For a detailed analysis of CFAR detectors, interested readers can refer to [9] and reference therein or to [14] and Chapter 16 in [15].

After detection of the frequency peak in down- and up-sweep spectrum, (5.5) is solved for computing target range and velocity.

5.2.3 Azimuth Direction of Arrival Estimation

The estimation of target DOA is not new in radar literature. The estimation problem, when only a single target is present, is well understood and approached in many books (see, for instance, [15] and references therein).

Traditionally, in automotive ACC radar, the azimuth DOA is estimated by the monopulse technique (or simultaneous lobing) [16]. Even if estimators that are more sophisticated have been proposed [17–21], real radar systems still use this simple technique.

The monopulse tracker is defined [16] as one in which information concerning the angular location of a target is obtained by comparison of the signal received in two or more simultaneous beams. In principle, it can work with just a single pulse (hence the name *monopulse*), but quite often multiple pulses are used to increase the angular resolution.

There are several methods by which a monopulse angle estimate can be made. The most popular and used in automotive radar is the amplitude-comparison monopulse, which compares the amplitudes of the signals received in two squinted fixed beams, taking both the sum Σ and the difference Δ of the two antenna patterns, as shown in Figures 5.8 and 5.9. In Figure 5.8 the two squinted antennas are plotted and the angular position θ_{TG} of the target is shown. In Figure 5.9 the two sum Σ and difference Δ antenna patterns are highlighted. In this figure, the two-way sum beamwidth is 3 degrees.

The sum channel is employed on transmission (with an approximate typical beamwidth of 12 degrees), while both sum and difference are used on reception. The estimate $\hat{\theta}_{TG}$ of the target DOA is a function of the ratio of the Δ and

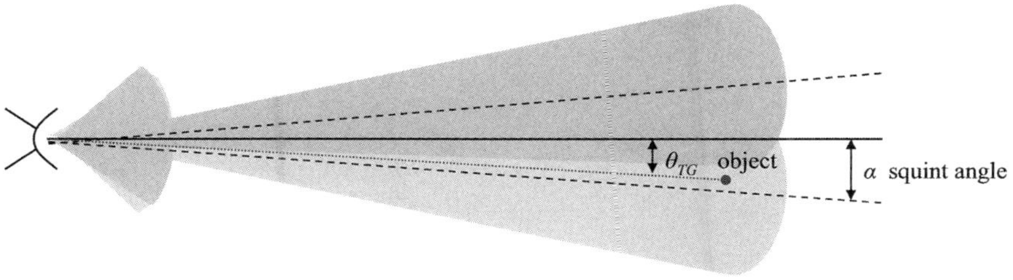

Figure 5.8 Squinted antenna beampatterns for monopulse technique.

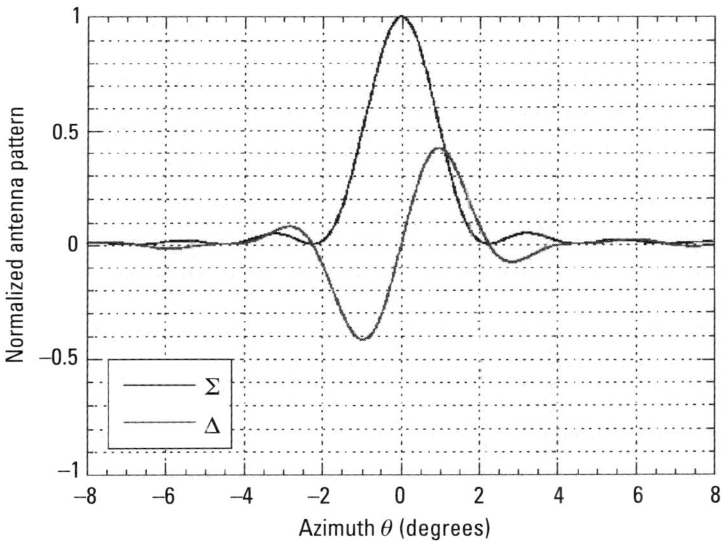

Figure 5.9 Example of normalized two-way antenna patter for Δ and Σ channels.

Σ channel outputs z_Δ and z_Σ. In detail, if a target has been detected, the amplitude of the two received signals will be used for the azimuth angle estimation. The signal processor forms the monopulse ratio defined by $r = \mathcal{Re}\{z_\Delta / z_\Sigma\}$, where $\mathcal{Re}\{\ \}$ denotes the "real part." In absence of disturbance and in presence of a single target, the monopulse ratio reduces to

$$r = \mathcal{Re}\left\{\frac{g_\Delta(\theta_{TG})}{g_\Sigma(\theta_{TG})}\right\} \tag{5.17}$$

where $g_\Delta(\cdot)$ is the difference pattern and $g_\Sigma(\cdot)$ is the sum pattern. Assuming that the ratio r is monotonic in off-boresight angle, the angular location of the target may be uniquely determined. It is evident from (5.17) that r does not depend on the amplitude of the target, so the monopulse estimate is not affected by the target fluctuations.

When the radar can use N samples $\mathbf{z}_\Delta = [z_\Delta(0) \quad \cdots \quad z_\Delta(N-1)]^T$ and $\mathbf{z}_\Sigma = [z_\Sigma(0) \quad \cdots \quad z_\Sigma(N-1)]^T$, the monopulse ratio and the estimate of the target DOA can be calculated for each sample, and finally, obtain an estimate as

$$\hat{\theta}_{TG} = \frac{1}{N}\sum_{n=0}^{N-1}\hat{\theta}_{TG}^{(n)} \tag{5.18}$$

where $\hat{\theta}_{TG}^{(n)}$ is the single-pulse target DOA estimate.

There are also more complex monopulse techniques that use more than two beams for increasing the field of view and the angular resolution and accuracy [22].

Another technique often used in automotive radar for angle estimation is the sequential lobing (SL). The measurement principle of the SL is similar to that of the monopulse technique.

In SL a single beam is switched between two (or more) squinted angular position (see Figure 5.10) to obtain an angle measurement. The difference in amplitude between the voltages in the two switched positions is a measure of the angular displacement of the target from the switching axis [16]. The advantage of the SL with respect to the monopulse is that only a receiver and signal

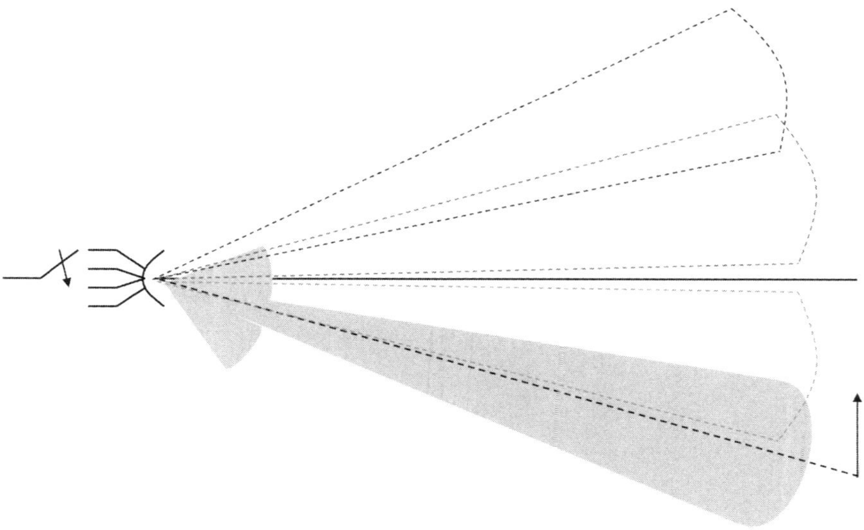

Figure 5.10 Sequential lobing.

processing unit is necessary (in the monopulse there are two, one per antenna); the disadvantage is that the angle estimation is sensitive to target fluctuations occurred during the switching.

The mechanical beam switching could be avoided by choosing a phased array antenna performing digital beamforming (DBF), where the beam scanning is induced by controlling the electrical phase difference between adjacent receiving (Rx) or transmission (Tx) antenna elements. Actually the DBF is able to form multiple receive beams by means of a digital processor, and it is more often considered when these multiple, simultaneous, and directive beams are necessary. Naturally, this involves complex antenna and millimeter-wave circuit configurations [17].

A typical circuit configuration of a phased array antenna is shown in Figure 5.11.

For use with a phased array antenna to detect the azimuth direction of the receiving signal, there are many methods with high azimuth resolution, such as the maximum likelihood (ML), multiple signal classification (MUSIC), and ESPRIT [23]. All these algorithms are computationally much heavier than monopulse or sequential lobing, but they can resolve more than one target at a time, with very high angular resolution [24] and without requiring very large antennas [25].

So, digital beam steering coupled with high-resolution DOA estimation algorithms can have a potential to overcome most limitations of today's

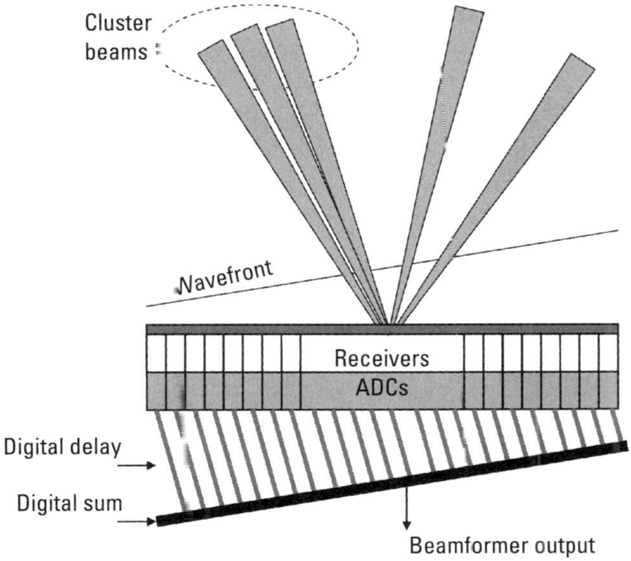

Figure 5.11 Phased array and digital beamforming.

automotive radar. However, with technology available today, the digital beam forming approach is still some years away from deployment on the mass market. Limiting factors are—among others—A/D converter resolution, A/D converter frequency bandwidth, and DSP computational power.

While for the azimuth angle the required resolution can be sometimes <2 degrees with wide field-of-view (>40 degrees), in the elevation (vertical) plane, there is usually no requirement for a wide angular coverage. A single beam antenna is typically used with sufficiently narrow beamwidth (4 degrees, for example) to differentiate between the vehicle ahead and the ground or the overhead bridges [3].

5.2.4 Target Tracking

After detecting the presence of a target, and after estimating its motion parameters, the radar must track the target itself in order to input the collision warning system whose ultimate goal is to reduce the risk of the vehicle accident. Therefore, tracking is one of the most important function of an automotive radar.

For the successful tracking of a moving target, it is essential to extract as much information as possible about the target state from the available observations. Therefore, good models describing the target dynamics and the sensor itself are necessary. As the information on the target's kinematics and sensor characteristics are generally known, most of the tracking algorithms base their performance on the a priori defined mathematical model of the target, which are assumed to be sufficiently accurate.

Since the target parameters measured by the radar are in polar coordinates (range, azimuth, range rate, or radial velocity) it may seem natural to perform the tracking in this coordinate system. Actually, most of the target motion models are much easier in Cartesian coordinates, so the tracking is done in the Cartesian system. The relations between Cartesian and polar coordinate systems are highly nonlinear, and the automotive radar tracking is a nonlinear filtering problem.

5.2.4.1 Nearly Constant Velocity Target Model

The nearly constant velocity target model is one of the most common dynamic target models used in automotive tracking. In this case the discrete-time equation of target motion in Cartesian coordinate system is as follows:

$$\mathbf{x}_{k+1} = \mathbf{F}\mathbf{x}_k + \mathbf{w}_k \tag{5.19}$$

where

$$\mathbf{x}_k = \begin{bmatrix} x_k & \dot{x}_k & y_k & \dot{y}_k \end{bmatrix}^T, \mathbf{F} = \begin{bmatrix} 1 & T_s & 0 & 0 \\ 0 & 1 & 0 & 0 \\ 0 & 0 & 1 & T_s \\ 0 & 0 & 0 & 1 \end{bmatrix} \tag{5.20}$$

T_s is the sampling time, and the state vector $\mathbf{x}_k$ consists of the position and the velocity of target. The process noise $\mathbf{w}_k$ is assumed to be zero mean Gaussian distributed with covariance matrix

$$\mathbf{Q} = E\left\{\mathbf{w}_k \mathbf{w}_k^T\right\} = q \begin{bmatrix} T_s^3/3 & T_s^2/2 & 0 & 0 \\ T_s^2/2 & T_s & 0 & 0 \\ 0 & 0 & T_s^3/3 & T_s^2/2 \\ 0 & 0 & T_s^2/2 & T_s \end{bmatrix} \tag{5.21}$$

and q is the process noise variance. This model is said to be "nearly constant velocity" since the acceleration along x and y axes are modeled as small white noise.

The tracking with much more complex target models is analyzed in [26–28].

Suppose that our measurement is linear in the target state; that is:

$$\mathbf{z}_k = \mathbf{H}\mathbf{x}_k + \mathbf{v}_k \tag{5.22}$$

where $\mathbf{z}_k = \begin{bmatrix} x_k^m & y_k^m \end{bmatrix}^T$ (the measured coordinates), $\mathbf{H} = \begin{bmatrix} 1 & 0 & 0 & 0 \\ 0 & 0 & 1 & 0 \end{bmatrix}$ and $\mathbf{v}_k = \begin{bmatrix} v_x & v_y \end{bmatrix}$. In (5.22), the observation noise $\mathbf{v}_k$ is assumed to be zero mean normally distributed with a known diagonal covariance matrix $\mathbf{R} = E\left\{\mathbf{v}_k \mathbf{v}_k^T\right\} = \begin{bmatrix} \sigma_x^2 & 0 \\ 0 & \sigma_y^2 \end{bmatrix}$.

With a linear model of the state equation and of the observation, with Gaussian process and observation noise, mutually independent, the optimum tracking filter is the Kalman one [29].

The prediction equation of the Kalman tracking algorithm provides the predicted target state $\mathbf{x}_{k+1}$ based on the observation until the previous instant k.

Prediction

$$\hat{\mathbf{x}}_{k+1|k} = \mathbf{F}\hat{\mathbf{x}}_{k|k} \tag{5.23}$$

The update equation provides the estimated target state at the instant k+1, based upon all the measurements until the instant k+1.

Update

$$\hat{\mathbf{x}}_{k+1|k+1} = \hat{\mathbf{x}}_{k+1|k} + \mathbf{K}_{k+1}\left(\mathbf{z}_{k+1} - \mathbf{H}\hat{\mathbf{x}}_{k+1|k}\right) \tag{5.24}$$

The predicted state covariance is

$$\mathbf{P}_{k+1|k} = \mathbf{Q} + \mathbf{F}\mathbf{P}_{k|k}\mathbf{F}^T \tag{5.25}$$

The updated state covariance is

$$\mathbf{P}_{k+1|k+1} = \mathbf{P}_{k+1|k} - \mathbf{K}_{k+1}\mathbf{S}_{k+1}\mathbf{K}_{k+1}^T \tag{5.26}$$

where $\mathbf{S}_{k+1}$ is the predicted measurement covariance:

$$\mathbf{S}_{k+1} = \mathbf{R} - \mathbf{H}\mathbf{P}_{k+1|k}\mathbf{H}^T \tag{5.27}$$

and $\mathbf{K}_{k+1}$ is the gain of the Kalman filter

$$\mathbf{K}_{k+1} = \mathbf{P}_{k+1|k}\mathbf{H}^T\mathbf{S}_{k+1}^{-1} \tag{5.28}$$

The Kalman filter is linear, and the matrices $\mathbf{P}_{k+1|k}$, $\mathbf{P}_{k+1|k+1}$, $\mathbf{S}_{k+1|k}$, and $\mathbf{K}_{k+1}$, can be calculated offline.

The steps of the Kalman filter are shown in Figure 5.12.

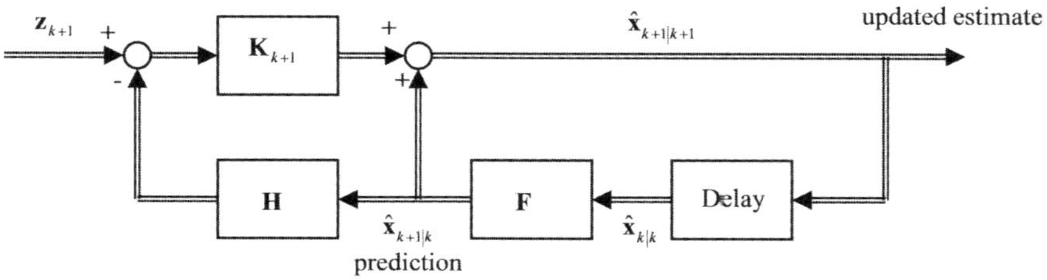

Figure 5.12 Kalman filter scheme.

Unfortunately, usually an automotive radar does not provide linear measures of the target state in Cartesian coordinates, but range r, azimuth angle θ, and range-rate $\dot{r}$ measurements, and the observation model is

$$\mathbf{z}_k = \mathbf{h}(\mathbf{x}_k) + \mathbf{v}_k \tag{5.29}$$

where $\mathbf{z}_k = \begin{bmatrix} r_k^m & \theta_k^m & \dot{r}_k^m \end{bmatrix}^T$, $\mathbf{h}(\mathbf{x}_k) = \begin{bmatrix} r_k & \theta_k & \dot{r}_k \end{bmatrix}$, and $\mathbf{v}_k = \begin{bmatrix} v_r & v_\theta & v_{\dot{r}} \end{bmatrix}$.

In (5.29), the observation noise is now assumed to be zero mean Gaussian distributed with diagonal covariance matrix $\mathbf{R} = E\left\{\mathbf{v}_k \mathbf{v}_k^T\right\} = diag\left\{\sigma_r^2, \sigma_\theta^2, \sigma_{\dot{r}}^2\right\}$.

We can observe that

$$\begin{aligned} r_k &= h_r(\mathbf{x}_k) = \sqrt{x_k^2 + y_k^2},\ \theta_k = h_\theta(\mathbf{x}_k) = \tan^{-1}\left(\frac{y_k}{x_k}\right) \\ \text{and } \dot{r}_k &= h_{\dot{r}}(\mathbf{x}_k) = \frac{x_k \dot{x}_k + y_k \dot{y}_k}{\sqrt{x_k^2 + y_k^2}} \end{aligned} \tag{5.30}$$

Since these measurement are a nonlinear function of the state, the Kalman filter cannot be used in the previous formulation. The extended Kalman filter (EKF) [30] is often used in its place. The EKF consists of the local linearization of the nonlinear measurement functions $\mathbf{h}(\mathbf{x}_k)$, such that

$$\mathbf{H}_k = \left[\nabla_{\mathbf{x}_k} \mathbf{h}^T(\mathbf{x}_k)\right]^T \Big|_{\mathbf{x}_k = \hat{\mathbf{x}}_{k|k-1}} \tag{5.31}$$

where $\left[\nabla_{\mathbf{x}_k} \mathbf{h}^T(\mathbf{x}_k)\right]^T$ is the Jacobian of $\mathbf{h}(\mathbf{x}_k)$.

In the EKF the two main equations are

Prediction

$$\hat{\mathbf{x}}_{k+1|k} = \mathbf{F}\hat{\mathbf{x}}_{k|k} \tag{5.32}$$

Update

$$\hat{\mathbf{x}}_{k+1|k+1} = \hat{\mathbf{x}}_{k+1|k} + \mathbf{K}_{k+1}\left(\mathbf{z}_{k+1} - \mathbf{h}_{k+1}\left(\hat{\mathbf{x}}_{k+1|k}\right)\right) \tag{5.33}$$

The predicted state covariance is $\mathbf{P}_{k+1|k} = \mathbf{Q} + \mathbf{F}\mathbf{P}_{k|k}\mathbf{F}^T$; the updated state covariance is $\mathbf{P}_{k+1|k+1} = \mathbf{P}_{k+1|k} - \mathbf{K}_{k+1}\mathbf{S}_{k+1}\mathbf{K}_{k+1}^T$, where $\mathbf{S}_{k+1}$ is the predicted measurement covariance $\mathbf{S}_{k+1} = \mathbf{R} - \mathbf{H}_{k+1}\mathbf{P}_{k+1|k}\mathbf{H}_k^T$ and $\mathbf{K}_{k+1}$ is the gain of the EKF $\mathbf{K}_{k+1} = \mathbf{P}_{k+1|k}\mathbf{H}_{k+1}^T\mathbf{S}_{k+1}^{-1}$.

The use of the Jacobian, the first order approximation of $\mathbf{h}(\mathbf{x}_k)$, sometimes can cause inconsistency of the covariance S_{k+1} when the linearization is poor, which results in improper filter gains $\mathbf{K}_{k+1}$ and then inconsistency of the filter estimates.

To overcome this problem, much more complex tracking filters have been proposed, such as the unscented Kalman filter (UKF) and the particle filter (PF) [31]. An excellent survey on the tracking techniques can be found in [30, 32].

5.3 SRR Radar

The main functionalities of the SSR are the parking aid, precrashing sensing, stop and go, and blind spot surveillance.

For the parking aid, the radar is intended to warn the driver if the distance between the vehicle and an obstacle or human being is below a critical value. Due to the very low speed of the car in parking, the radar need low update rates (100 ms), but a wide angular range in azimuth and a very accurate measure of the distance to the nearest object.

For stop and go functionality, the radar should warn for cut-in collisions in order to enable reduction of speed in time to avoid contact. In very dense traffic situations, this application reduces a large amount of accidents. The update rate must be high (50–100 Hz). Distance measurement parameters are similar to those of an ACC system. For correct distance control, a wide range in velocity has to be covered with good precision, as well as a wide area in azimuth angle.

For precrashing sensing, the radar is intended to help the driver to react very quickly and to activate all necessary components (brakes or steering) in the car to avoid an accident or at least minimize consequences of an impact with reduction of the vehicle velocity. Due to this reason, the update rate must be very high (200 Hz).

The requirements for the different SRR functions are summarized in Table 5.3.

Since the required range resolution is generally very high (on the order of cm), the SRR uses ultrawideband technology; that is, the transmitted signal bandwidth exceeds "the lesser of 500 MHz or 20% of the center frequency," according to the FCC definition [33]. This transmit wideband can be obtained in different ways: either using nonsinusoidal pulses of a typical width of less than

Table 5.3
Requirements for the Different SRR Functions

	Parking Aid	Stop and Go	Precrash
Cycle time (ms)	100	10-20	5
Distance (m)			
Range	0.05-5	0.5-20	0.5-20
Accuracy	0.05	0.5	0.5
Resolution	—	1	1
Relative velocity (km/h)			
Range	—	−180 +180	−180 +180
Accuracy	—	1	1
Resolution	—	5	5
Azimuth angle (°)			
Angle of view	−90 +90	−60 +60	−60 + 60
Accuracy	5	2	2
Resolution	—	5	5

1.5 ns, and thus a bandwidth over 1 GHz, or using longer but phase-coded pulses and pulse compression.

In the first case, different type of signals respect this band requirement as Gaussian monocycle pulse, Gaussian doublet pulse (plotted in Figure 5.13), or Gegenbauer functions and Hermite functions [33].

The drawback of these waveforms is that, for increasing the probability of detecting the obstacles, the transmitted peak power can be very high, while, for complying with the FCC rules, the average equivalent isotropically radiated power (EIRP) emitted by the SRR UWB sensor must stay below –41.3dBm/MHz in the 22-GHz to 29-GHz band, and peak EIRP should be below 0 dBm/50 MHz in the same frequency band.

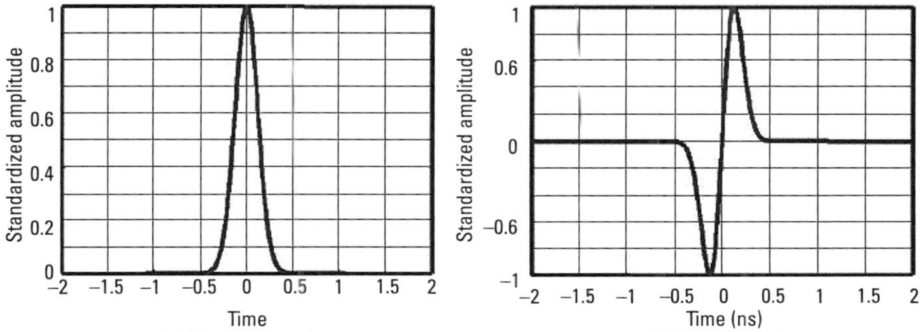

(a) Monocycle pulse (b) Gaussian pulse

Figure 5.13 Waveform of UWB SRR, Gaussian pulse, and Gaussian doublet.

Often to meet these stringent requirements on the EIRP and peak EIRP, coded pulses are used. They are longer pulses, phase modulated with M-bits code that reach the resolution requirements after pulse compression (or matched filtering). This is a traditional technique for enhancing radar performance and may be used to improve sensitivity for detecting targets at short and long range. If the transmit pulse with pulse width T and M-bits code is selected at a certain range, the compression gain associated with the binary phase codes is equal to M, and the peak value is M times larger than the peak power of the long pulse itself [34].

The general expression of the complex envelop of a phase-coded pulse is given by

$$u(t) = \frac{1}{\sqrt{T}} \sum_{m=1}^{M} u_m rect\left[\frac{t-(m-1)T_b}{T_b}\right] \tag{5.34}$$

where T_b is the duration of each subpulse or bit, $u_m = \exp(j\phi m)$ and the set of M phases $\{\phi_1, \phi_2, \ldots, \phi_M\}$ in the phase code. Flexibility exists with the choice of the code, allowing designers to trade various performance parameters, with the limitation that dynamic range is often restricted by code auto-correlation properties (see Chapter 20 in [15], as well as [34]. In UWB radar the most common waveform is the binary phase modulated Barker code, whose maximum known length is M=13.

The received signal (autocorrelation), after pulse compression, is plotted in Figure 5.14 for M=13.

As for the pulsed radar, the distance between the radar (on the car) and the obstacle is measured based upon the delay of the received pulse with respect to the transmitted one. CFAR techniques can be used in both range and frequency domain for increasing the probability of detection and improving the range and frequency estimation as already shown for the LRR.

Unlike the LRR, the SRR has a very wide field of view, and, in this case, the monopulse azimuth estimation with only two beams can have some problems. For example, when driving down a narrow street with infrastructure objects symmetrically present on both sides of the street, the objects fall in the same range and frequency bins, and thus the wide angle monopulse radar cannot resolve them as separate objects. The result is that these targets are inaccurately placed by the radar directly in the path of the car.

One more difference between LRR and SRR is that, while LRR does not need precise information on target elevation, SRR does because it must distinguish, for instance, between lower-lying objects (soda cans, water bottles, or other road debris) over which a vehicle could safely drive and taller obstacles of interest (vehicles, humans, or chain link fences) over which a vehicle must not

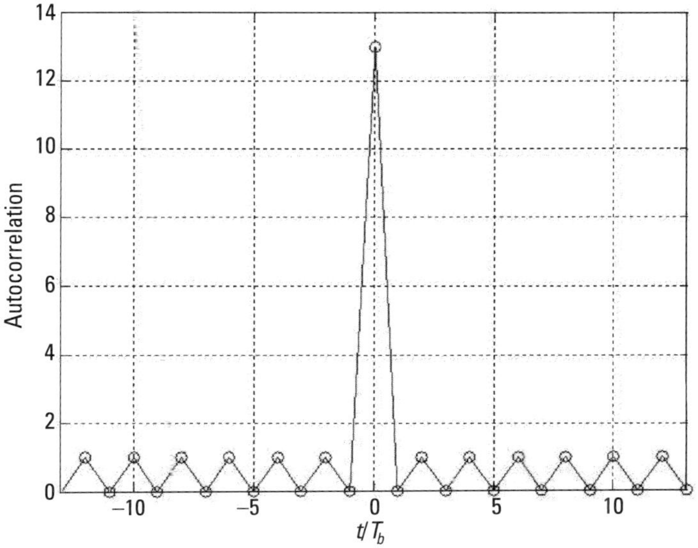

Figure 5.14 Autocorrelation function of phase-coded radar pulse with 13-element Barker code.

drive. To solve this problem, a possible approach could be to use the monopulse technique in elevation as well that is able to characterize low-profile targets on the road surface as obstacles not of interest, while still correctly reporting the average height for vertically distributed objects that must be avoided.

5.4 Conclusions

The aim of this chapter was to introduce the main statistical signal processing techniques used by a single automotive radar for detection and tracking of a target at long and short distances.

The research activity on this topic is far from being definitive and final. There is an increasing interest in mounting more precise and less costly radars even on low-class cars, possibly networked with other sensors like cameras and lidars. Moreover, the idea of using multiple radars on the same vehicle is becoming real [35], and more work is needed to optimize network performance and cost.

References

[1] Hoare, E. G., and R. Hill, *System Requirements for Automotive Radar Antennas*, London, UK: Institution of Electrical Engineers.

[2] Corredoura, P., Z. Baharav, B. Tabwe, and G. Lee, "Millimeter-Wave Imaging System for Personnel Screening: Scanning 10^7 Points a Second and Using No Moving Parts," *SPIE Conf. on Passive Millimeter Wave Imaging Technology IX*, Vol. 6211, 2006.

[3] Abou-Jaoude, R., "ACC Radar Sensor Technology, Test Requirements, and Test Solutions," *IEEE Trans. on Intelligent Transportation Systems*, Vol. 4, No. 3, September 2003, pp. 115–122.

[4] Mizutani, K., and R. Kohno, "Inter-Vehicle Spread Spectrum Communication and Ranging System with Concatenated EOE Sequence," *IEEE Trans. Intell. Transport. Syst.*, Vol. 2, No. 4, Dec. 2001, pp. 180–191.

[5] Rohling, H., and M. Kronauge, "Continuos Waveforms for Automotive Radar Systems," Ch. 7 in *Waveform Design and Diversity for Advanced Radar Systems*, edited by F. Gini, A. De Maio, and L. Patton, IET Radar, Sonar and Navigation Series, Vol. 22, 2012.

[6] Kok, D., and J. S. Fu, "Signal Processing for Automotive Radar," *2005 IEEE International Radar Conference*, pp. 842–846.

[7] Winkler, V., "Range Doppler Detection for Automotive FMCW Radars," *2007 European Radar Conference*, pp. 166–169.

[8] Oppenheim, A. V., and R. W. Schafer, *Discrete-Time Signal Processing*, Englewood Cliffs, NJ: Prentice Hall International, 1989.

[9] Gandhi, P. P., and S. A. Kassam, "Analysis of CFAR Processors in Nonhomogeneous Background," *IEEE Trans. on Aerospace and Electronic Systems*, Vol. 24, No. 4, July 1988, pp. 427–445.

[10] Ozgunes, L., P. P. Gandhi, and S. A. Kassam, "A Variably Trimmed Mean CFAR Radar Detector," *IEEE Trans. on Aerospace and Electronic Systems*, Vol. 28, No. 4, October 1992, pp. 1002–1014.

[11] Hansen, V. G., and J. H. Sawyers, "Detectability Loss Due to Greatest of Selection in a Cell-Averaging CFAR," *IEEE Trans. on Aerospace and Electronic Systems*, AES-16, January 1980, pp. 115–118.

[12] Trunk, G. V., "Range Resolution of Targets Using Automatic Detectors," *IEEE Trans. on Aerospace and Electronic Systems*, AES14, September 1978, pp. 750–755.

[13] Rohling, H., "Radar CFAR Thresholding in Clutter and Multiple Target Situations," *IEEE Trans. on Aerospace and Electronic Systems*, AES-19, July 1983, pp. 608–621.

[14] Richards, M. A., *Fundamentals of Radar Signal Processing*, New York: McGraw-Hill, 2005.

[15] Richards, M. A., J. A. Scheer, and W. A. Holm, *Principles of Modern Radar: Basic Principles, Vol. I*, Raleigh, NC: Scitech Publishing, 2010.

[17] Tokoro, S., K. Kuroda, A. Kawakubo, and K. Fujita, et al., "Electronically Scanned Millimiter-Wave Radar for Pre-Crash Safety and Adaptive Cruise Control System," in *Intelligent Vehicles Symposium*, Jun. 2003, pp. 304–309.

[18] Wenig, P., M. Schoor, O. Gunther, and Y. Bin, et al., "System Design of a 77 GHz Automotive Radar Sensor with Superresolution DOA Estimation," *2007 International Symposium on Signals, Systems and Electronics*, pp. 537–540.

[19] Farina, A., F. Gini and M. Greco, "Multiple Target DOA Estimation by Exploiting the Amplitude Modulation Induced by Antenna Scanning," *IEEE Transaction on Aerospace and Electronic Systems*, Vol. 38, No. 4, October 2002, pp. 1275–1286.

[20] Greco, M., F. Gini, and A. Farina, "Asymptotic Maximum Likelihood DOA Estimation of Multiple Radar Targets," *Signal Processing Elsevier*, Vol. 84, No. 6, June 2004, pp. 969–979.

[21] Greco, M., F. Gini, A. Farina, and L. Timmoneri, "Direction-of-Arrival Estimation in Radar Systems: Moving Window and Approximate Maximum Likelihood Estimator," *IET Radar, Sonar and Navigation*, Vol.3, No. 5, 2009, p. 552–557.

[22] Alland, S. W., "Antenna Requirements and Architecture Tradeoffs for an Automotive Forward Looking Radar," *1998 IEEE Radar Conference*, pp. 367–372.

[23] Schoor, M., et al., "High-Resolution Angle Estimation for an Automotive FMCW Radar Sensor," in *IRS*, Cologne, Germany, Sep. 2007.

[24] Stoica, P., and R. Moses, *Introduction to Spectral Analysis*, Upper Saddle River, NJ: Prentice Hall, 1997.

[25] Richter, M. D., M. Schneider, P. Wenig, and R. Weigel, "77 GHz Automotive Digital Beamforming Radar with SiGe Chipset," *2010 German Microwave Conference*, pp. 210–213.

[26] Altendorfer, R., "Observable Dynamics and Coordinate Systems for Automotive Target Tracking," *2009 IEEE Intelligent Vehicles Symposium*, 2009, pp. 741–746.

[27] Eidehall, A., and F. Gustafsson, "Combined Road Prediction and Target Tracking in Collision Avoidance," *2004 IEEE Intelligent Vehicles Symposium*, pp. 619–624.

[28] Li, X. R., "Survey of Maneuvering Target Tracking. Part I: Dynamic Models," *IEEE Trans. on Aerospace and Electronic Systems*, Vol. 36, No. 4, October 2003, pp. 1333–-1348.

[29] Kay, S. M., *Fundamentals of Statistical Signal Processing: Estimation Theory*, Upper Saddle River, NJ: Prentice Hall PTR, 1993.

[30] Bar-Shalom, Y., X. Rong Li, and T. Kirubarayan, *Estimation with Applications to Tracking and Navigation*, New York: John Willey & Sons, 2001.

[31] Gustafsson, F., F. Gunnarsson, N. Bergman, and U. Forssell, et al., "Particle Filters for Positioning, Navigation and Tracking," *IEEE Transactions on Signal Processing*, Special Issue on Monte Carlo Methods for Statistical Signal Processing, Vol. 50, No. 2, pp. 425–437.

[32] Ristic, B., S. Arulampalam, and N. Gordon, *Beyond the Kalman Filter: Particle Filters for Tracking Applications*, Norwood, MA: Artech House, 2004.

[33] See http://www.fcc.gov.

[34] N. Levanon, E. Mozeson, *Radar Signals*, Hoboken, NJ: Wiley-Interscience, 2004.

[35] F. Folster, H. Rohling, "Data Association and Tracking for Automotive Radar Networks," *IEEE Trans. on Intelligent Transportation Systems*, Vol. 6, No.4, 2005, pp. 370–377.

6

Low-Power Radar Front-End for E-Health and Harbor Surveillance: Implementation Examples

6.1 Summary

In this chapter, two case studies of low-power/low-cost radar front-end for specific applications are presented, starting from the problem description up to the design, simulation, and realization of the prototypes. The emphasis is on the radio front-end circuits whose design and simulation has become much easier than in the past thanks to modern simulation tools and accurate high-frequency device modeling. At the same time, the unitary cost of new generation integrated radar sensors has been reduced by orders of magnitude thanks to the utilization of modern integrated circuit technologies. The front-end contains all of the radiofrequency circuits and subsystems necessary to generate the waveform to be transmitted, together with the output and input RF filters, low noise amplifier, mixer, and so on. The output of the radar receiver can be AD converted, and digital elaboration can be accomplished by purposely designed integrated digital signal processor or by traditional elaboration tools running on an external elaboration system, typically a PC. Although the emphasis is mainly on circuits and subsystems, some basic information will be given about the antennas and the digital signal processing used to test the prototypes.

The first case we present is a miniaturized pulsed E-health (EH) radar for noninvasive continuous monitoring of heart and breath activities, which does not require digital elaboration. The second one is a low-power radar front-end for harbor surveillance (HS). In this second case, the output of the radio front-end receiver is sampled and elaborated by an external system even if, in the future, a dedicated DSP could be embedded in a multichip module together with the radio interface for a more compact, cheaper, and less power consuming solution.

For the design and realization of EH radar, an entirely integrated solution was chosen. It was possible to foresee a large-scale production of the device, whose main utilization could be the monitoring of cardiopathic patients during normal daily activities or the monitoring of breathing activity of sleeping adults or infants affected by recurrent apneas. It is well known that, for large production scales, more suited solutions consist of a fully integrated system on a single chip: a feasibility study and design are presented in this chapter, whereas related technologies have been described and discussed in Chapter 2.

For the design and realization of HS radar, instead, a small-scale production has been foreseen: some tens of devices for each harbor, thousands of pieces in total. For this reason, a hybrid solution was chosen—discrete devices mounted on a substrate in which passive devices (matching networks, filters, power divider and combiners, and so on) are integrated. This solution does not require the utilization of sophisticated technological processes and, consequently, access to a silicon foundry. It can be realized in any electronic laboratory equipped with a photolithographic etching process and a work station for bonding. Hybrid circuits for RF and microwave applications, commonly called microwave integrated circuits (MICs), represent the best solution for small- and medium-scale production and for prototype realization. Section 6.3 of this chapter is dedicated to a concise description of the hybrid technology for MIC design and realization.

6.2 Miniaturized Radar for E-Health

A portable radar system can allow the monitoring in a contactless way of some vital functions such as the heart activity and the respiratory one. In fact, the movement of the heart walls, as well as the slower one of the chest due to breathing, interact with the transmitted electromagnetic wave and modulate the back scattered one. Both CW radar, using the Doppler effect, and pulsed radar, measuring the change of distance between the radar sensor and the target, can be used. In the case of respiratory activity, which is characterized by slow movements in a frequency range about five times lower than cardiac frequency, the pulsed radar is more effective. In fact, the shift Doppler Δf depends on the target velocity V_R, on the frequency f_0, and on the propagation velocity c of the transmitted wave through the relationship:

$$\Delta f = 2\frac{v_R}{c} f_0 \tag{6.1}$$

Equation (6.1), for $f_0 = 10$ GHz, an estimated chest velocity v_R of a few centimeters per second, gives a value of Δf in the range of a few hertz (for c the value of the EM wave phase velocity in vacuum/air has been used). An output signal in this frequency range is very difficult to be monitored mainly due to (1) the low frequency fluctuations of the amplifier offset, and (2) the flicker noise affecting the operations of any electronic devices at low frequencies. These causes of disturbances cannot be eliminated by using a low pass filter in that they are located in the same frequency range of the signal. For this reason, a pulsed radar solution has been preferred.

In the past, several realizations have been presented of single chip (or chipset) implementing the radar for biomedical applications [1–8]: we will refer here to that described in [6, 7].

The block scheme of the system is described in Figure 6.1 in which a UWB pulsed radar solution is presented [6, 7].

The term UWB applies in case the frequency band occupied by the signal is at least 500 MHz or more than 20% of the central frequency. To this kind of application (UWB modulation), a frequency range starting from 3.1 up to 10.6 GHz has been dedicated by international organizations (FCC for the United States, and ETSI for Europe). This frequency range can be utilized without any authorization (unlicensed band) provided that the effective isotropic radiated power spectral density of the transmitted signal is lower than -41.3 dBm/MHz.

The transmitted pulse is partially reflected at each interface it passes through: air/skin, muscle/bone, bone/abdominal cavity, abdominal cavity/heart wall, and so on. In this way, an echo is generated at each interface: for each transmitted pulse several echoes with different delays are back scattered to the

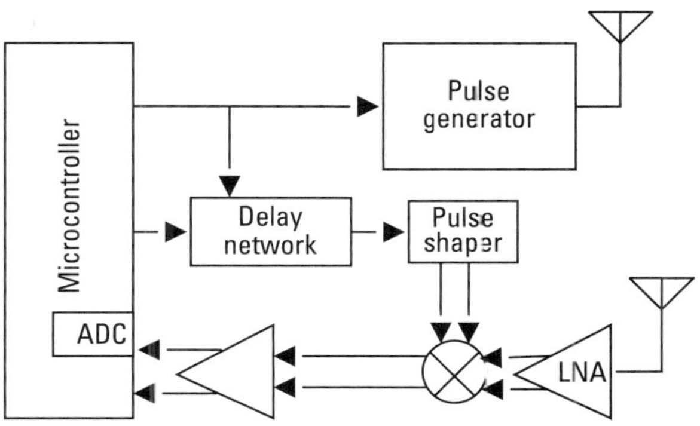

Figure 6.1 Block diagram of a pulsed radar for biomedical applications [7].

receiving antenna. By choosing a suited value of the delay introduced by the delay block in Figure 6.1, the replica of the transmitted pulse generated by the shaper arrives at the multiplier input together with the echo coming from the selected interface. The output of the multiplier is different from zero just during the time interval in which delayed and reflected pulses are superimposed at the multiplier input. So the output of the integrator (a low-pass filter) attains its maximum when the pulses perfectly coincide, whereas it goes to zero when they do not superimpose at all. The small cyclic movement of the interface causes a cyclic variation of the reciprocal delay between the replica of the pulse generated by the shaper and the echo received by the receiving antenna, causing a cyclic variation of the output of the integrator. For a continuous monitoring of the instantaneous displacement $\Delta x(t)$ of the interface from a reference position, the pulse duration ΔT should be equal to twice the time the EM wave takes to cover a distance equal to the maximum (Δx_{MAX}) of Δx (i.e., $\Delta T = 2\ \Delta x_{MAX}/c$). In fact, if the delay time is set in such a way that when Δx=0, delayed and reflected pulses completely superimpose each other at the multiplier input, then when $\Delta x = \Delta x_{MAX}$, their reciprocal delay at the multiplier input will be ΔT, and the time interval in which they superimpose will be nulled. However, ΔT values larger than $2\Delta x_{MAX}/c$ can be used, provided that a reduction of the dynamic of the signal at the output of the integrator can be tolerated. In fact, in this case a partial superimposition between reflected and delayed pulses is present even when $\Delta x = \Delta x_{MAX}$. For instance, if ΔT=$4\Delta x_{MAX}/c$, when Δx=0 the two pulses are completely coincident at the multiplier input, whereas when $\Delta x = \Delta x_{MAX}$ they are superimposed for a time equal to $2\Delta x_{MAX}/c$ (i.e., ΔT/2). In this case, the output of the integrator will change between V_{MAX} (Δx=0) and V_{MAX} /2 ($\Delta x = \Delta x_{MAX}$) instead of between V_{MAX} and zero. This discussion shows that values of the pulse duration slightly larger than twice the time the EM wave takes to cover a distance equal to the maximum (Δx_{MAX}) of Δx can be used without a drastic lack of sensitivity of the radar sensor. With reference to the breath monitoring, we can suppose $\Delta x_{MAX} = 3$ cm at which a value of ΔT=2 $\Delta x_{MAX}/c = 200$ ps would correspond. For the calculus, the light velocity in air has been used for c: this fact implies that the interface to be monitored moves in air (breath monitoring); otherwise, the proper phase velocity in the medium should be used. In [9] the relative permittivity of some biological tissues are reported: the typical values at 10 GHz are around 10, corresponding to a propagation velocity three times (or more) slower than in air. A value of ΔT slightly higher than 200 ps is obtained in the case of heart monitoring; in fact, the cardiac walls displacement is around 1.5 cm, but the propagation velocity is reduced by about 1/3. In the prototype of the EH radar, which has been realized and tested, variable ΔT between 200 and 400 ps has been chosen in order to relax the specifications on the pulse generator (PG) design.

A preliminary feasibility study has demonstrated the possibility of realizing the entire transmitter/receiver section in a single chip. By supposing a power supply voltage of 1.2V, a differential output voltage of 900 mV at the transmitter output, a differential signal of 650 mV at the transmitting antenna input (a voltage drop of 0.25V due to the cables has been considered), and a differential antenna impedance of 100Ω, the transmitted power during the pulse, having a sinusoidal shape, is 0.528 mW. In order to estimate the available power at the input of the receiving antenna, the attenuation between transmitted and received pulse must be calculated. A rough estimation, confirmed also by other similar cases presented in the literature ([6, 7] and references therein), indicates, for heart monitoring, a value of the attenuation of about 80 dB due to path loss, to the reflection coefficient smaller than 1 at the interfaces and to the attenuation inside the biological tissues. It must be noted that, in the case of heart monitoring, the transmitting antenna is put in contact with the chest, so that far field equations for propagation and attenuation of EM wave cannot be used and near field equations have to be used. Details about this modeling are given in [6]. The attenuation is much smaller in the case of breath monitoring in that the antennas are put a few tens of centimeters from the chest and the interface to be monitored (i.e., between air and skin), so that no attenuation inside the biological tissues has to be considered. In the worst case (80 dB attenuation), a value of the reflected available power at the receiving antenna input $P_{AIN}=5.28\times10^{-12}$ W can be hypothesized. This value corresponds, at T_0=290K, to a signal-to-noise ratio (SNR_{OUT}) at the output of the LNA, for a single pulse, given by the following relationship:

$$SNR_{OUT} = \frac{P_{AIN}}{kT_0BN_F} \tag{6.2}$$

where N_F is the LNA noise figure, B=7.5 GHz (from 3.1 to 10.5 GHz) is the LNA band for UWB signal, and k is the Boltzmann constant. The denominator of (6.2) should be substituted by $kB\left[T_0\left(N_F-1\right)+T_i\right]$ if the absolute temperature at which the input of the amplifier is terminated was Ti. By using a prudential value of 10 dB for N_F, T_i= 298 K (25°C) (breath monitoring), (6.2) gives a value of SNR_{OUT} of about −18 dB. The situation would appear just a bit worse for heart monitoring: in this case T_i= 37°C, which is the mean temperature of human body. This value of SNR_{OUT} could appear too small for a correct detection of the back-scattered pulse at the receiver output. At this point, it must be considered that the pulses are repeated with a pulse rate PR between 1 and 40 MHz and that the output of the multiplier can be integrated for a given time T_{INT}. The choice of the integration time must be done by taking into account two different limitations. In fact, T_{INT} has to be quite long to maximize the signal-to-noise ratio without infringing upon the conditions for time resolution required for monitoring the phenomenon under observation. The period of the phenomena under observation is between 0.5s (heart) and 10s (breath). So an integration time of about 1/100 of the period of the fastest phenomenon under observation can be considered suited for a continuous monitoring with a sufficient time resolution. By using T_{INT}=5ms (both the heart and the chest displacement can be considered negligible in 5 ms) and PR=2 MHz, 10,000 quite identical and fully correlated pulses occur in T_{INT}. By integrating the output of the multiplier with a time constant equal to T_{INT}, an improvement of 40 dB (10Log10,000) of SNR_{OUT} is obtained. Consequently, the resultant value of SNR_{OUT} rises up from −18 to +22 dB, which can be considered well suited for a detailed analysis of the phenomenon under observation. This quite rough estimation demonstrates, at a first approach, the possibility to realize in a fully integrated way the radar sensor. In fact, the pulse duration (200–400 ps), the output power (0.528 mW with a power supply voltage of 1.2V), and the receiver noise figure (less than 10 dB) can be obtained by using any well-consolidated technology available on the market. The fully integrated radar sensor has been designed, simulated, and realized by using the 90 nm CMOS technology by STMicroelectronics. After the chip realization, the single blocks of the prototype have been characterized by means of a purposely designed test bench.

The schematics of the building blocks of the front-end [7] are shown in Figure 6.2. A differential topology has been chosen starting from pulse generator output up to antennas and, in the receiving section, for LNA, multiplier and integrator. A square wave generator with a maximum frequency P_R= 10 MHz is integrated in order to generate a triangular wave which, in turn, is shaped by a shaping network (SN in Figure 6.2) whose output is a bipolar Gaussian-like pulse. This shape of the pulse has been chosen to comply with the regulation requirement for UWB communication in terms of power spectral density mask. The output of the shaper is connected by a short differential cable to the transmitting antenna.

The receiving antenna, with the same characteristics of the transmitting one, receives the back scattered echoes that are pulses similar to the transmitted ones, even if a given level of linear distortions has to be taken into account due to the effect of the crossing of biological tissues and the reflection at the interfaces. The antenna gain used for testing was a slotted antenna with a gain of 10 dB realized on a soft substrate [10], whereas a purposely designed antenna, for use in respiratory activity monitoring, realized in FR4 has been described in [11].

The LNA topology is in three stages: common gate followed by two common source differential stages. The main characteristics of LNA are: gain = 22.7 dB @ 5 GHz, return loss S_{11}<−10.5 dB, NF= 6.5 dB, 1dB compression point $1dB_{CP}$= −19.7 dBm, and dissipated power P_D= 34.8 mW. A P-MOS Gilbert

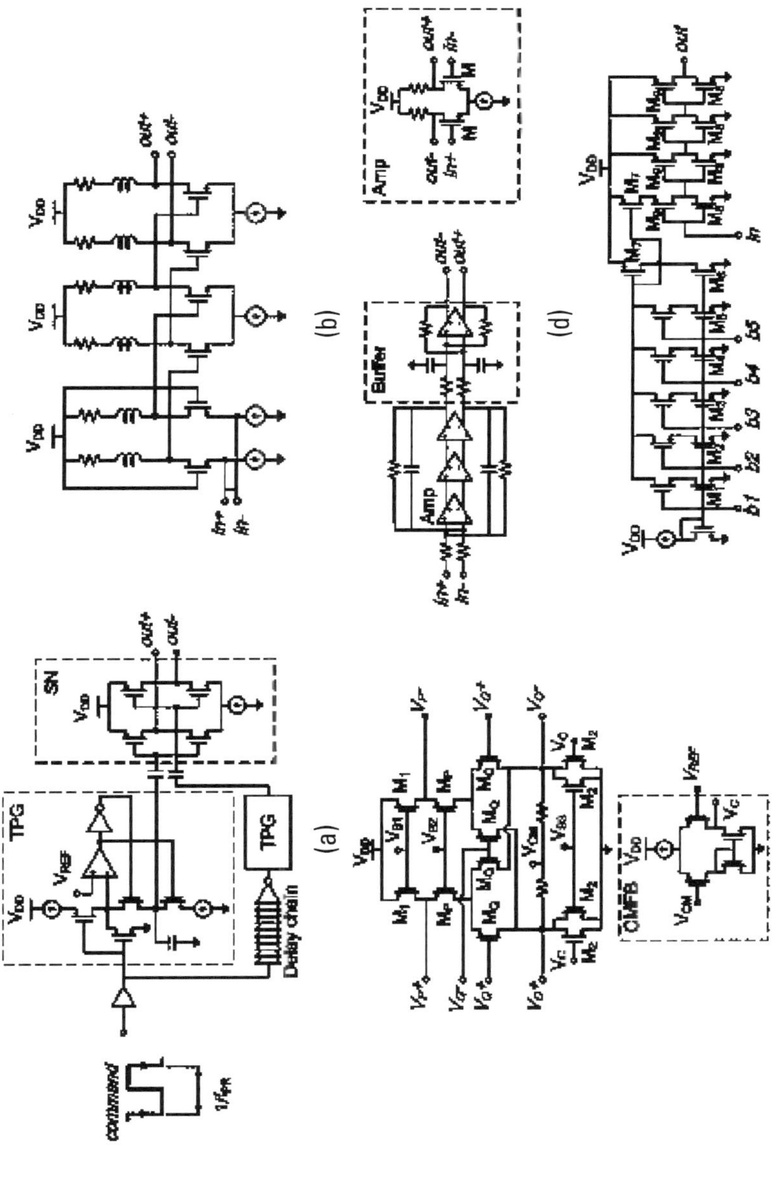

Figure 6.2 Schematic of the building blocks of EH radar front-end [7]: (a) pulse generator, (b) LNA, (c) multiplier, (d) integrator, and (e) delay generator; VDD=1.2V.

cell implements the multiplier whose power consumption is 3.7 mW. The multiplier output is sent to the differential integrator, which is implemented by means of a simple RC low pass filter preceded and followed by two stages that introduce a total voltage gain of 58 dB.

In Figure 6.3 the micrograph of the radar test chip is shown: the occupied area on silicon die is 1.5×1.3 mm^2. For testing, the die has been mounted on a QFN32 package whose external dimensions are 5×5 mm^2, whereas the antennas, whose dimensions are in the range of some centimeters depending on the required gain, could be realized in the future on a polymeric flexible substrate and "embedded" together with the radar microchip in a T-shirt, realizing the first prototype of a "wearable radar." The characteristic of the integrated EH radar front-end, together with the small value of the total power consumption (only 73.2 mW), makes the EH radar well suited to be "worn" by cardiopathic patients who could be equipped with an RF interface (e.g., Bluetooth or Zigbee) for data transmission to a base station or to a storage device for continuous, noninvasive monitoring during their normal activities.

A prototype of EH radar front-end has been successfully tested for the contactless monitoring of breathing activity in patients affected by nocturnal apneas and for the monitoring of babies during their sleep in order to prevent the phenomenon of crib deaths [7].

6.3 Microwave Integrated Circuit

The term MIC refers to RF and microwave circuits realized by integrating both passive and active devices in the same planar substrate. For passive devices (resistances, capacitors, and inductors), two possibilities are available: (1) they can be realized directly on the substrate, and (2) they can be purchased in packaged or in chip version and bonded on the substrate. For active devices, just the second possibility is available since it is not possible to realize them directly on the

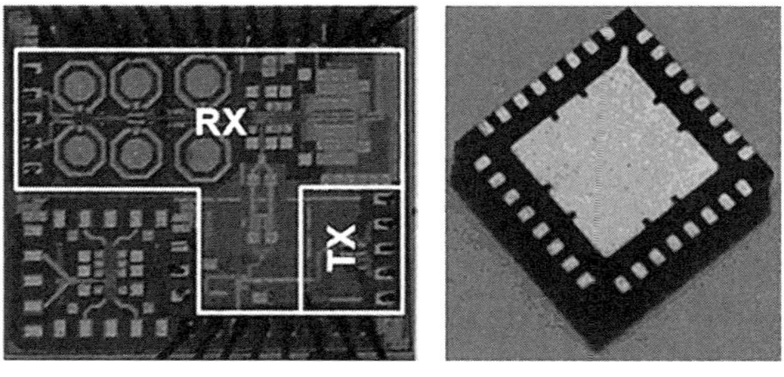

(a) (b)

Figure 6.3 (a) Micrograph of the EH radar front-end chip and (b) of the QFN32 in which it has been mounted [7]. The die size is 1.5 × 1.3 mm2.

substrate. For this reason, it would be more appropriate to use the term hybrid circuits for MIC, since this technology makes use of devices directly realized on the substrate together with discrete devices previously realized elsewhere and then bonded on the substrate.

MICs are composed of (1) the substrate, (2) the passive devices, and (3) the active devices. To these three components, both the connectors and the case must be added, which are essential to realize the link with other blocks of the system and to give a suitable shielding effect. The latter functionality is very important in order to avoid cross talk between different blocks, to reduce the effect of interferences and external noise, and to guarantee the stability of the active stages.

In Figure 6.4, a typical MIC realization is shown: the upper case has been removed to show the planar substrate with active and passive devices. In the figure, some typical distributed passive devices realized with a photolithographic process are also shown. A detailed analysis of theory and application of planar technology can be found, for instance, in [12].

Unlike MICs, MMIC are the true integrated circuits for RF and microwave applications in that they are entirely realized by specialized silicon foundry on the same substrate (active, passive, and interconnections). MMIC are realized on Si, GaAs, or other suited substrates by using modern technologies for integrated circuits that allow the realization of entire cells, blocks, and sometimes entire systems on the same chip. With respect to MIC, MMIC allow better performances in terms of maximum operating frequency, dimensions, reliability, and power efficiency, but their utilization can be considered just for large-scale production. In fact, it must be taken into account the fact that

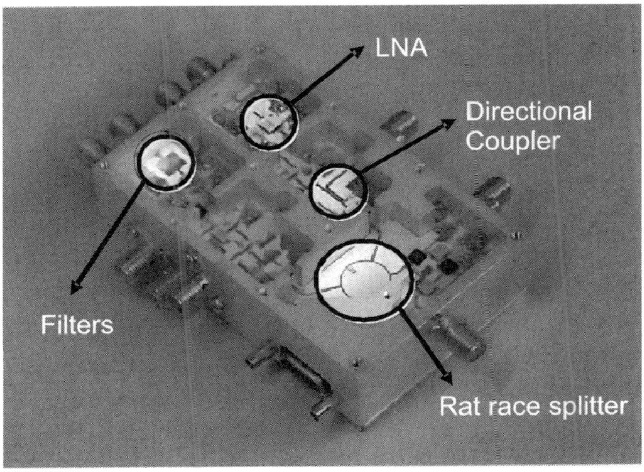

Figure 6.4 A hybrid circuit realized by using a soft substrate and enclosed in a purposely designed case with input, output, and test connectors.

an integrated circuit requires the preliminary realization of the masks and the availability of a suited technological process. The cost of the masks and the access to the technological facilities makes advantageous the utilization of MMIC technology just in case of a large-scale production (hundreds of thousand of pieces or more). Moreover, it must be considered that the possibility of realizing distributed elements on MMIC (transmission lines, matching networks, filters, and power combiners and dividers) is restricted to the higher frequency range of interest for microwave. In fact, the dimensions of distributed devices are comparable with the wavelength so that their realization on a chip is made possible just above a few tens of GHz, (i.e., for wavelength in the range of millimeters).

6.3.1 The Substrates

Two types of substrates are normally used for MIC: soft and hard. Soft substrates make use of materials similar to Teflon (Diclad, Duroid, and so on). The relative permittivity is between 2 and 3, and they are light, quite flexible, and cheaper than hard substrates. Soft substrates are mainly used in the frequency range between a few GHz and a few tens of GHz, up to the millimeter-wave range. Above this limit, hard substrates have to be used because the dielectric losses in soft substrates increase together with parasitic and spurious effects, which cause a performance degradation in terms of gain and power dissipation.

Hard substrates make use of high permittivity material, such as alumina (AL_2O_3) or berillium oxide. The relative permittivity is around 10 or more: this fact has a positive effect on the dimensions of distributed elements in that the wavelength is reduced by a factor of 3 or more, with respect to air; therefore, the dimensions of transmission lines, matching networks, filters, and so on are reduced by the same factor. By using hard substrates, it is possible to realize MIC in the frequency range up to 100 GHz or more.

Both soft and hard substrates are constituted by a dielectric sheet with a double metallization, in the top and down surface. The down surface metallization is used as ground plane, whereas in the upper metallization layer suited geometries can be realized by using standard photolithographic process. In this way, different passive devices, such as power combiners and dividers, directional couplers, spiral inductors, planar transformer and balun, interdigitized capacitors, and so on, can be realized [12].

Furthermore, by using a little more complex technological processes than the lithographic one, the deposition of thin metal film, ceramic pastes, and insulating layers is made possible, thus allowing the realization on the substrate of passive devices, such as resistors, metal-insulator-metal (MIM) capacitors, and multilayer inductors and transformers. Finally, it must be pointed out that in planar MIC (regardless of the substrate type) different typologies of planar antennas (dipole, patch, slotted, and so on, both active and passive) can be realized, as well as high directivity antenna arrays with or without electron beam forming. In other words, with MIC technology an entire microwave transmitter-receiver system, included antennas, LNA, and power amplifiers can be designed, simulated, and realized *in house* by using quite low-cost instrumentation and facilities.

6.3.2 Design, Simulation, and Realization of Microwave Integrated Circuits

Several CAD tools are available for the design and simulation of MIC, but the most complete and widespread one is the Advanced Design System (ADS) by Agilent Technologies. ADS makes available a very complete integrated simulation tool starting from the bottom level of the electromagnetic simulation of circuits, antennas, and case, up to the top level of the entire communication system simulation.

Furthermore, a specific ADS module makes available integration with Cadence, the most widespread CAD tool for integrated circuits design. In this way, the design kit of a particular technological process for integrated circuits production (e.g., 65 nm CMOS by STMicroelectronics) can be used in an ADS environment for design and simulation, and then the final circuit design can be transferred to the silicon foundry for large-scale production. In other words, by using ADS, both MIC and MMIC can be designed and simulated. In Figure 6.5 a typical screenshot of ADS is shown.

After the simulation, when the results comply with the initial specifications, the layout of the circuits is extracted, and the board with passive distributed devices and the slots for passive and active discrete devices is realized.

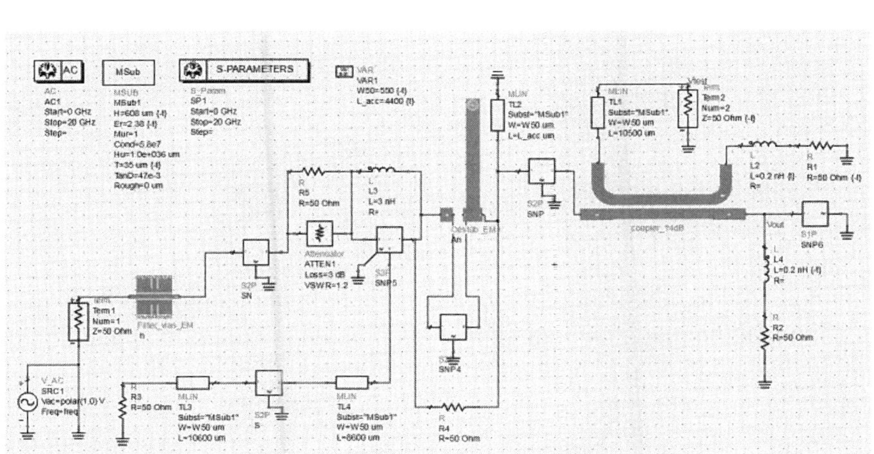

Figure 6.5 ADS screenshot of the receiver section of a radar Doppler.

6.4 Low-Cost Radar Prototype for Harbor Surveillance

The target of this project was the design, simulation, realization, and testing of a low-power (a few watts), low-cost (less than $1,000 for radio front-end) radar, capable of monitoring the presence and tracking the movements of any kind of ship inside and in the neighborhood of a harbor. The radar sensor should be able to make 3D images of ships and vessels with a resolution of about 0.5m. To this end, a single transmitting antenna and three receiving channels connected to three antennas in spatial diversity, in order to allow interferometric analysis, will be used. The way in which 3D reconstruction can be accomplished is beyond the scope of this book, so we will examine in the following the single channel radar front-end (i.e., a radar with only one receiving antenna). The overall design of the harbor surveillance system is based on a wireless sensor network whose nodes are single radar sensors distributed in the harbor area, allowing the data fusion at network level of the image coming from all of the nodes. The entire network could be constituted by some tens of radar sensors, which should be characterized by low dissipated power, small dimensions, and, most of all, a low level of EM pollution.

An FMCW solution is the most suited for such an application, as it is capable of giving information both on the distance R and on the radial velocity V_R of the target by using quite low values of the output power (a few watts) for some kilometers of coverage range.

The typical block diagram of an FMCW radar front-end is shown in Figure 6.6.

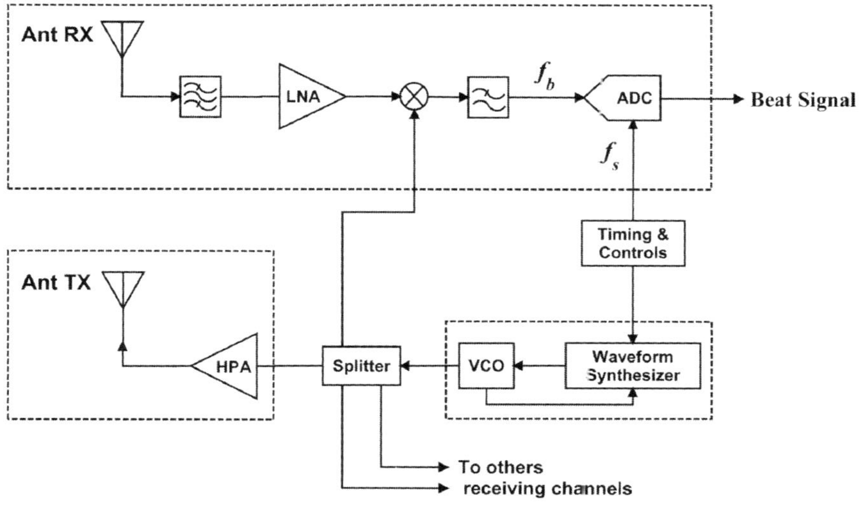

Figure 6.6 Block diagram of a FMCW radar front-end.

6.4.1 Feasibility Study and Dimensioning

The principle of operation of the HS radar can be summarized as follows. The instantaneous frequency transmitted by the output antenna is $f_T(t) = f_{MIN} + At$. When a target at a distance d reflects the transmitted wave, the back-scattered wave arrives at the receiving antenna with a delay τ = 2d/c. The received frequency at instant t is $f_R(t) = f_{MIN} + A(t-\tau)$. At the output of the mixer at time t, and we have a frequency f_{IF} equal to the difference between f_T and f_R. That is,

$$f_{IF} = A\tau = A2d/dc \qquad (6.3)$$

By measuring f_{IF} the value of d can be evaluated provided that A is known.

The specifications are shown in Table 6.1.

The FM band depends on the required resolution through the well-known relationship:

$$B = \frac{c}{2d_R} \qquad (6.4)$$

where $B = f_{MAX} - f_{MIN}$ is the difference between the maximum (f_{MAX}) and minimum (f_{MIN}) instantaneous value of the frequency, c is the light velocity in air, and d_R is the distance (or range) resolution. Equation (6.4) for the required resolution of 0.5m gives a value of 300 MHz; therefore, the values of f_{MIN} and f_{MAX} have been chosen equal to 10.5 and 10.8 GHz.

The behavior of the instantaneous frequency of the transmitted constant wave is shown in Figure 6.7, where T_{SW} is the rise time (i.e., the time required for the instantaneous frequency to rise from its minimum value f_{MIN} to its maximum value f_{MAX}). During a time interval T_{SW} the instantaneous frequency f rises up with a velocity A, where A is measured in s^{-2}. The instantaneous frequency is generated by a voltage-controlled oscillator (VCO) and inserted in a

Table 6.1

Frequency	X band (8–12.5 GHz)
Waveform	FMCW
Antenna	60° azimuth
Range	> 1.5 km
Range resolution	0.5m
Receiver dynamics	> 96 dB
SNR at the receiver output	SNR ≥ 20 dB
Channel number	3 (for 3D imaging)
Maximum Output power	10W

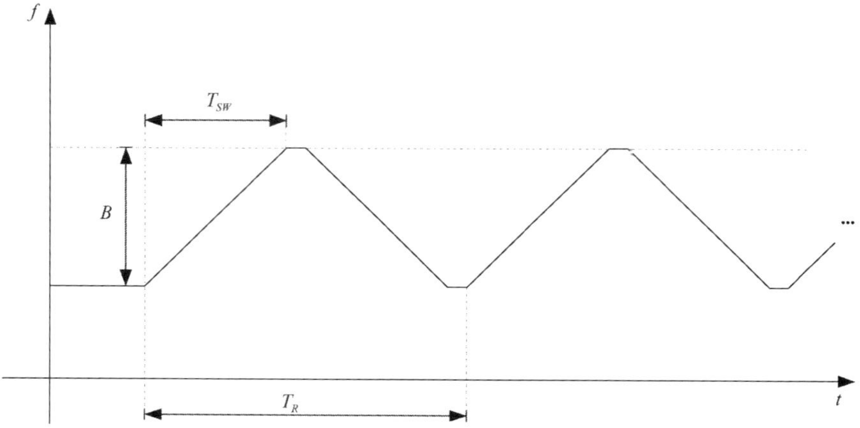

Figure 6.7 Instantaneous frequency versus time.

phase locked loop (PLL), which can be digitally controlled by a microcontroller (μC). The μC send a trigger signal at the beginning of each new frequency ramp to the DSP in order to allow synchronization. It is worth emphasizing that the frequency reference of the PLL is a crystal oscillator that continuously oscillates at its resonance frequency, so that the phase coherence of the generated FMCW over different ramps is guaranteed by the synchronization with the same reference oscillation. The coherence accuracy depends only on the phase noise of the reference source. Maximum values of A for commercial PLL, for frequencies around 10 GHz, are in the range of $10^{12}\ s^{-2}$ (e.g., see [13]). For higher values of A, the VCO linearity is no longer guaranteed, and the generated instantaneous frequency shows a more complicated dependence on time t: this has to be avoided for a correct and simple application of FMCW principles of operation.

As far as the SNR at the receiver output is concerned, it depends on the following quantities: the output power P_{CW} of the transmitter , the noise figure N_F of the receiver, the cross radar section σ of the target, the transmitting and receiving antenna gain (G_{TX} and G_{RX}, respectively), the total losses L in air and inside the system, the maximum distance R_p of the target, the equivalent available total noise power spectral density N_0=KTN_F, at the input of the receiver (for T=290K), and the bandwidth B_F of the filter at the output of the mixer receiver.

$$SNR = \frac{P_{RX}}{kTB_F N_F} = \frac{P_{CW}\lambda^2 G^2}{(4\pi)^3} \frac{1}{L} \frac{\sigma}{R_p^4} \frac{1}{kTB_F N_F} \tag{6.5}$$

In (6.5) P_{RX} represents the received power, whereas G_{TX} and G_{RX} are both supposed to be equal to G.

The output signal of the mixer, after amplification and filtering, is sampled and transferred to the digital domain for digital signal processing (DSP). A discrete Fourier transform (DFT) is calculated on the sampled data, whose frequency resolution Δf is equal to the inverse of the time record length ΔT: $\Delta f = 1/\Delta T$. In the digital domain, the equivalent of the filter bandwidth B_F is the frequency resolution of the DFT, so that B_F has to be substituted by Δf in (6.5). In our case the time record length ΔT coincides with T_{SW} (see Figure 6.7). Some details about the processing of the beat signal will be given in Section 6.4.2.

As in the case of EH radar presented in Section 6.2, for the HS radar the SNR can be improved by a factor N, where N is the number of identical repetitions of the received signal, by integrating the output for a time $T_{INT} = N\,T_{SW} = N/\Delta f$ (i.e., $N = T_{INT}\Delta f$). The coherence, as already stated, is guaranteed by the reference oscillator of the PLL.

The resulting value of the SNR after DSP with an integration time equal to T_{INT}, SNR_{OUT}, is given by

$$SNR_{OUT} = \frac{P_{CW}\lambda^2 G^2}{(4\pi)^3} \frac{1}{L} \frac{\sigma}{R_D^4} \frac{T_{INT}}{kTN_F} \tag{6.6}$$

SNR_{OUT} is shown in Figure 6.8 for different values of T_{int}. The value of the different parameters used for the evaluation is represented in the inset in the bottom of the figure, whereas the total loss L is the combined effect of air loss (L_{am}) and system loss (L_{sys}).

A commercial horn antenna with 13 dBi gain has been used for the prototype testing, but a low-cost patch antenna array (patent pending) realized by using a soft substrate will be used for the final version of HS radar. Due to the low values of the target velocity (less than 50 km/h), in a time interval of 40 ms, the target displacement is less than 0.55m, very similar to the required range resolution. Therefore, an integration time of 40 ms does not impair in a significant way the accuracy in the target movement's reconstruction. This means that a value of SNR_{OUT} of 20 dB can be obtained with an output power equal to 2W (see Figure 6.8). Obviously, the positive effect of integration on is effective just in the case in which the coherence can be guaranteed (see Section 6.4.2 for this aspect).

Until now, the effect of the relative velocity of the target with respect to the radar sensor (i.e., the Doppler effect) has been neglected. Due to this effect, in the case of a target moving toward the radar at a radial velocity v_R, the

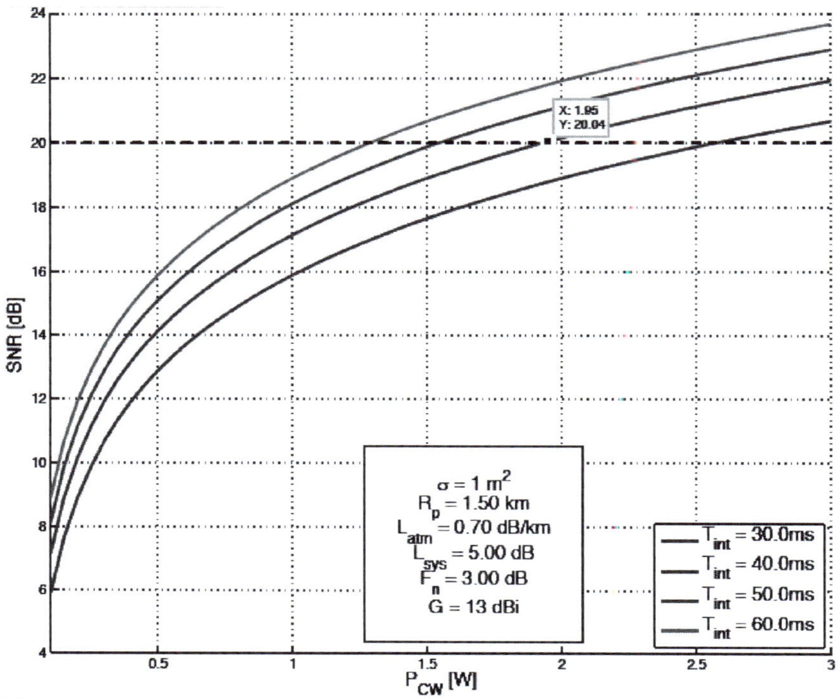

Figure 6.8 Signal-to-noise ratio after DSP for different values of integration time TINT.

received frequency is shifted by a quantity Δf given by (6.1). It is worth noting that Δf is lower than 1 kHz for $v_R \leq 50$ km/h.

It is possible to show that, due to Doppler effect, in case of moving targets, the frequency f_{IF} at the mixer output is shifted by $-\Delta f$ in the half period in which the instantaneous frequency increases and by $+\Delta f$ in the other half period; see (6.3). This means that the information on the velocity and on the distance of the target are both contained in the beat signal (see Figure 6.7) at the output of the radar front-end. An example of the way in which they can be extracted by means of DSP is presented in Section 5.2.1 of Chapter 5 for automotive FMCW radar. For HS radar a different choice has been made: some details will be given in Section 6.4.3.

6.4.2 Realization

By taking into account the results of the feasibility study in Section 6.4.1 and the availability of discrete and integrated devices on the market, the specifications shown in Table 6.2 have been carried out for the circuit blocks shown in Figure 6.7.

Table 6.2
Specifications for the Circuit Blocks in Figure 6.7

Receiver NF (included ADC)	≤ 5 dB
LNA gain	29 dB
Mixer conversion loss	8 dB
Minimum beat frequency (d=50m)	200 KHz
Maximum beat frequency (d=1250m)	5 MHz
IF band	100 KHz–5 MHz
IF gain	43 dB

A maximum output power of 2W is obtained due to the stage marked as a high-power amplifier (HPA) in Figure 6.7. A hybrid PLL has been utilized with a VCO in the range of 10–12 GHz, capable of generating a FMCW with a maximum value of A in the required range of $10^{12}s^{-2}$ (a value of $6\times10^{11}s^{-2}$ has been chosen corresponding to T_{SW}=500 μs in order to obtained the required value of B=300 MHz).

The frequency range of the beat signal depends on the maximum distance that has to be covered. For harbor surveillance, by considering also the fact that the complete coverage is obtained by installing a wireless radar sensor network, a suited coverage range for each sensor is 1.5 km. For d=1.5 km, (6.3) gives f_{IF}= 6 MHz. A sixth-order low pass Chebyshev filter, with 6-MHz bandwidth, has been used as an antialiasing filter before sampling the beat signal at a sampling rate of 12.5 MSa/s.

The prototype of the system has been realized by using the hybrid technology described in Section 6.3. With reference to Figure 6.6, the dotted lines circumscribe the part of the system contained in a single board; each board has been mounted in a custom designed case with microwave connectors in order to allow individual testing before proceeding to a more compact single board realization of the prototype.

Several simulations have been carried out at both the circuit and system levels by using ADS (see Section 6.3); an example regarding the results of the simulation of the receiver stage (LNA + input filter) is presented in Figure 6.9, where the parameters S21 (insertion gain) and S11 (return loss) are shown.

In Figure 6.10 the receiver section board is shown. Receiver and transmitter sections have been realized in separated boards to make easier the prototype test: they will be contained in the same compact shielded case in the final realization. The case will be connected to the antennas (receiving and transmitting) and to the power supply, whereas the low frequency output of the mixer, after low pass filtering and amplification (test port in Figure 6.10), will be made available for sampling and DSP.

For a single sample of the radar front-end shown in Figure 6.10, the cost of the devices (active, passive, discrete, and integrated) as well as the realization of the board and of the case can be estimated around US$500 (€370). It is evident that in the case of series production of several hundreds or thousands of samples, the unit cost would drop in a significant way (at least 50%).

6.4.3 Data Processing

In this section, some information will be given about the processing of the signal at the output of the radar front-end receiver.

After AD conversion at 12.5 MS/s with 14-bit resolution, the signal is ready for digital signal processing. For SH radar prototype testing, DSP has been implemented in a Matlab platform running on a PC. In the final version of the radar, DSP will be performed on a dedicated board, but the processing steps will be unchanged and the final result will be the identification of targets inside the operating range of the radar in terms of distance and radial velocity.

The processing algorithm is different from that described in Section 5.2.1 of Chapter 5 for automotive FMCW radar. In fact, in order to detect radial velocities as low as 1 kn (1 nautical mile per hour), according to (6.1), the frequency resolution of FFT should be less than 36 Hz, and, consequently, the time record length should be greater than 27.7 ms. This means that the number of samples for the FFT, at 12.5 MHz of sampling rate, should be greater than

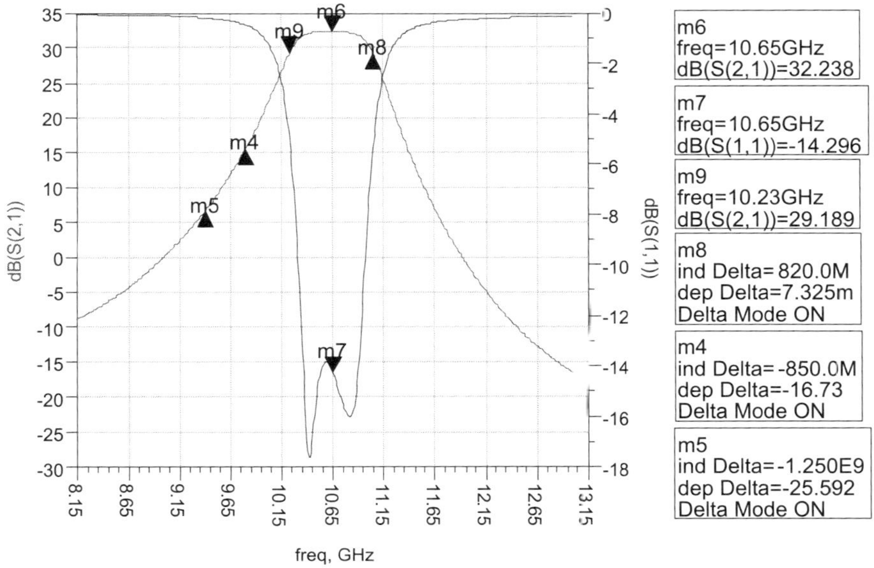

Figure 6.9 Insertion gain (S21) and return loss (S11) of the receiver section of the HS radar front-end.

Figure 6.10 The prototype of the receiver of the HS radar.

347,222 (i.e., 524,288), which is the closest power of 2. This would require the utilization of purposely designed and very expensive DSP for real-time elaboration. For this reason, the processing algorithm, that has been implemented is different: it consists of a two-dimensional (2D) FFT, shortly described in the following paragraph.

First of all, the coherent period of integration (CPI) has to be determined. CPI corresponds to the T_{INT} in (6.6). The serial flux of samples in a CPI, taken at a sampling period equal to 80 ns (the inverse of the sampling rate) is organized in a matrix with L rows and N columns. Each column contains the samples corresponding to a single up or down frequency ramp whose duration is T_{SW} (see Figure 6.7). The number of columns N is equal to CPI/T_{SW}, whereas the number of rows L is equal to the number of samples in T_{SW}. For instance, using $T_{SW} = 500\ \mu s$ and CPI = 50 ms results in N=100 and L=6250. In this way, two adjacent positions in the same column correspond to consecutive samples of the beat signal (i.e., two samples taken at 80 ns each other), whereas two adjacent positions in the same row correspond to samples of the beat signal taken at T_{SW} seconds of each other. At this point, after suited correction, to take into account the fact that the range component of the beat frequency reverses sign when the direction of sweep is reversed, windowing and zero padding, a 2D FFT is performed on the matrix. A complete discussion about this topic is beyond the aim of this book and can be found, for instance, in [14]. Here, it is worth just to say that the result of 2D FFT, after proper elaboration, can be organized in a 3D map in which the radial velocity and the distance of a given target are represented in the X and Y axes, respectively, and the "intensity" of the corresponding echo is represented in Z axis (or in a 2D graph, by using a false color map).

The laboratory preliminary tests, hardware, and data processing showed that the performances of the HS radar match the required characteristics for radio front-end, but to verify the true resolution and sensitivity of the whole radar sensor network, a campaign of on-field measurements has to be carried out in the future.

Acknowledgments

The author wishes to thank Dr. R. Massini, who designed, realized, and tested the complete prototype of HS radar front-end, and Dr. S. Lischi (Ph.D., University of Pisa) who implemented DSP algorithms.

References

[1] Fanucci, L., S. Saponara, T. Bacchillone, and M. Donati, et al., "Sensing Devices and Sensor Signal Processing for Remote Monitoring of Vital Signs in CHF Patients," *IEEE Transactions on Instrumentation and Measurements*, 2013

[2] Staderini, E. M., "UWB RADAR in Medicine," *IEEE Aerospace and Electronic System Magazine*, Vol. 17, Jan. 2002, pp. 13–18.

[3] Lin, J. C., "Microwave Sensing and Physiological Movement and Volume Change: A Review," *Bioelectromagnetics*, Vol. 13, 1992, pp. 557–565.

[4] McEwan, T. E., "Body Monitoring and Imaging Apparatus Method," US Patent 5 573 012, Nov. 12, 1996.

[5] I. J. Immoreev and S. V. Samkov, "Ultra-Wideband RADAR for Remote Measuring of Main Parameters of Patient's Vital Activity," *Radio Physic and Radio Astronomy*, Vol. 7, 2002, pp. 404–407.

[6] Zito, D., D. Pepe, B. Neri, and D. De Rossi, "Feasibility Study of a Low-Cost System-on-a-Chip UWB Pulse RADAR on Silicon for Heart Monitoring," *IEEE Waveform Diversity and Design Conference*, 2007, pp. 32–36.

[7] Zito, D., et al. "SoC CMOS UWB Pulse RADAR Sensor for contactless Respiratory Rate Monitoring" , IEEE Trans. Biomedical System and Circuits. Vol. 5, 2011.

[8] Neri, B., and S. Saponara, "Advances in Technologies, Architectures and Applications of Highly Integrated Low-Power Radars," *IEEE Aerospace and Electronic Systems Magazine*, Vol. 27, No. 1, 2012, p. 25.

[9] Gabriel, C., and S. Gabriel, "Compilation of the Dielectric Properties of Body Tissues at RF and Microwave Frequencies," 2000, http://www.brooks.af.mil/AFRL/HED/hedr/reports/dielectric/home.html.

[10] Zito, D., D. Pepe, M. Mincica, and F. Zito, et al., "Wearable System-on-a-Chip UWB Radar for Contact-Less Cardiopulmonary Monitoring: Present Status," *Proc. 30th Annual International IEEE EMBS Conference*, Vancouver, BC, August 20–24, 2008, pp. 5274–5277.

[11] Pepe, D., and D. Zito, "Planar Differential Antenna for UWB Pulse Radar Sensor," *COMCAS 2013*, IEEE Conference Publications, pp. 1–4.

[12] Pozar, D. M., *Microwave Engineering*, 4th edition, Wiley, 2013.

[13] Hittite, IC PLL HMC70xxxxx series, www.hittite.com.

[14] Jankiraman, M., 'Design of Multi-Frequency CW Radars," *SciTech Pub.,* 2007.

7

Automotive Radar IC Design: 24-GHz UWB and 77-GHz FMCW Implementation Examples

This chapter presents meaningful examples of automotive radar IC design. The aim is to provide the reader with a circuit-level focus and a better knowledge of millimeter-wave design issues addressed to an important application such as the automotive radar. For the sake of completeness, both 24-GHz SRR and 77-GHz LRR applications are considered in Sections 7.2 to 7.5. Special attention will be given to the description of the selected architectures, circuit topologies of key building blocks, design procedures, and experimental results. The focus is put on the circuit-level design, as signal processing for radar was discussed in Chapter 5. All the circuits presented in this chapter are integrated in a SiGe HBT BiCMOS technologies by STMicroelectronics [1, 2], featuring high-speed npn transistors and optimized back-end-of-line (BEOL) for high-quality passives.

Section 7.1 is fully dedicated to a brief description of adopted BiCMOS technology platforms that are among the best silicon processes for millimeter-wave applications and thus well represent the state of the art.

Section 7.2 describes the design and the experimental performance of the first reported fully integrated UWB SRR sensor for automotive applications at 24 GHz. Sections 7.3 to 7.5 put the focus on the implementation of millimeter-wave transceivers for 77-GHz LRR.

7.1 Silicon Technologies for Automotive Radar

The main challenge for a widespread and pervasive use of automotive driver assistance systems (ADASs) even in low-end cars is a drastic reduction of radar sensor cost, which may enable a mass-market production. A crucial step has been replacing bulky and expensive discrete electronics with MMICs, including millimeter-wave, analog, and digital blocks in a chip-set implementation or better in a fully integrated solution. Although millimeter-wave circuits have traditionally been exclusive domain of GaAs, InP, and pHEMT technologies, research and industrial efforts in the past decade demonstrated the full maturity of low-cost silicon-based technologies that nowadays are well suited for millimeter-wave radar. In this scenario, silicon manufacturers addressed their effort toward the development of HBT SiGe bipolar technologies with f_{max} around 200 GHz [2–5], often including sub-μm CMOS transistors to enable digital circuitry integration [2–4]. Another important feature of such millimeter-wave advanced platforms is a high-performance BEOL, which provides top copper metals and thick intermetal oxides to lower parallel/series ohmic losses and parasitic capacitance of passive components [6] (i.e., capacitors, inductors, transformers, and microstrips). At the present time, SiGe-based processes represent the optimum choice for both 24-GHz [7–10] and 77-GHz radar [11–18], thanks to the high level of transmitted power, the good linearity/noise performance, the wide operative temperature range (i.e., –40 to +125° C), and the high reliability. Indeed, all these requirements are mandatory to meet automotive radar applications. As far as pure CMOS is concerned, several implementations of 24-GHz and 77-GHz radar front-ends have been reported in past years [19–25]. A fully integrated CMOS automotive radar is expected for the near future [26], especially for short-range applications at 24 GHz that require less aggressive output power levels. Nevertheless, SiGe BiCMOS technologies still have a significant advantage over their radio frequency (RF) CMOS counterpart.

The implementation examples described in Sections 7.2 to 7.5 are developed by making use of two successive generations of a 0.13-μm SiGe HBT BiCMOS platform by STMicroelectronics. The first technology, namely BiCMOS9 [1], features 166-/175-GHz f_T/f_{max} npn transistors, dual-V_T dual-gate-oxide CMOS devices, MIM and metal-oxide-metal (MOM) capacitors, accumulation MOS (A-MOS) varactors, and a 6-level metal copper back-end. This technology is well suited to implement K-band ICs. It was profitably used for the 24-GHz radar sensor described in Section 7.2. Moreover, despite the relatively low f_T/f_{max}, the adoption of proper circuit arrangements and optimized passive design allowed pushing the technology up to W-band implementations [12, 14, 27], as described in Section 7.3.

An advanced SiGe BiCMOS platform, namely BiCMOS9MW [2], was instead exploited to implement the 77-GHz transmitter and receiver (TX/RX) radar front-ends described in Sections 7.4 and 7.5, respectively. This technology features very-high-speed SiGe HBTs (i.e., 230-/280-GHz f_T/f_{max}), along with a millimeter-wave optimized BEOL for high-quality integrated passives. Presently, it is among the few silicon-based technologies capable of implementing W-band ICs with fully complied performance. For the sake of completeness, a comparison between BiCMOS9 and BiCMOS9MW is provided in Figure 7.1.

Besides the speed of the bipolar transistors, the main differences between the two BiCMOS technologies are in the thickness of the Cu top metals (i.e., 3 μm) and oxide between metals. Important is also the distance of the metals from the substrate that sets the value of bulk parasitic capacitances. Figure 7.1 shows how a technology advance has to improve all these aspects to enhance performance, which means higher operating frequency, higher linearity, lower noise, better isolation, and so on.

7.2 A Fully Integrated 24-GHz UWB SRR Sensor

The implementation of a complete radar sensor on a single chip is still very challenging and involves many issues such as proper system architecture, efficient detection methodology, robust radio front-end structure, and customized local digital signal processing (DSP). In this scenario, UWB sensors based on RX analog correlation represent an attractive solution for a cost-effective automotive SRR [28]. This section describes the first published fully integrated 24-GHz UWB radar sensor, which was implemented in a 0.13-μm SiGe BiCMOS process [1, 7]. The sensor integrates on the same die as the RX/TX front-

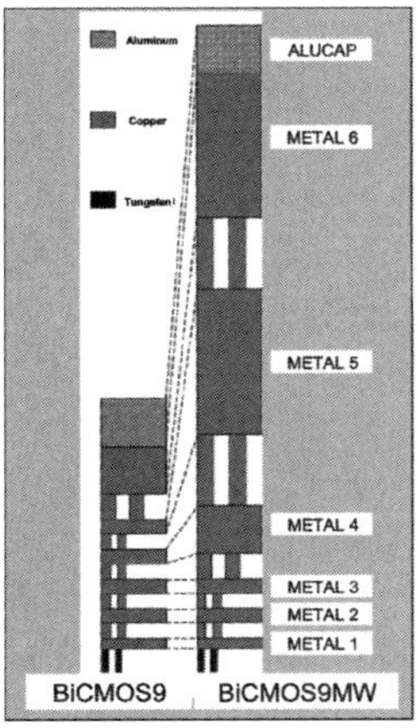

	BiCMOS9	BiCMOS9MW
SiGe HBT	160/175-GHz f_T/f_{max}	230/280-GHz f_T/f_{max}
npn BV_{CEO}	1.8 V	1.6 V
CMOS node	130 nm (1.2 / 2.5 V)	130 nm (1.2 / 2.5 V)
BEOL	6 Cu metals (1 thick) + ALUCAP	6 Cu metals (3 thick) + ALUCAP
Capacitors	MIM / MOM	MIM / MOM
Varactors	MOS / Diode	MOS / Diode
Spiral inductors	YES	YES

Figure 7.1 Comparison between BiCMO9 and BiCMOS9MW technologies.

end, along with both baseband analog and digital sections. It operates with standard supply voltages (i.e., 2.5V and 1.2V for the RF/analog and digital sections, respectively).

7.2.1 Sensor Architecture

Pulsed correlation radar architecture shows some advantages in comparison with FMCW radars. Indeed, pulsed radar allows detection of both target distance and its relative velocity with better accuracy, higher resolution, and less average power consumption. Moreover, circuit complexity due to the frequency modulation is avoided. The operating principle of a UWB radar sensor based on RX analog correlation scheme is depicted in the simplified block diagram in Figure 7.2.

A CW source is pulsed by switch SW_TX and driven by the baseband pulse generator PG, thus producing the UWB transmitted signal (a). This is a pulse train with a proper pulse width, T_{pulse}, and pulse repetition period, T_{PR}, which is received back by the sensor after the time of flight (TOF). At the RX section, an on-chip delayed replica of the transmitted pulse is generated by switch SW_RX and used as LO for the homodyne down-conversion, whereas the RX input is the echo signal reflected by the target. Correlation occurs when reflected pulses and delayed pulses overlap, which means that the applied delay nearly equals the TOF. Target detection requires that a threshold is crossed by the sensor output. Of course, analog integration of the down-converted signal

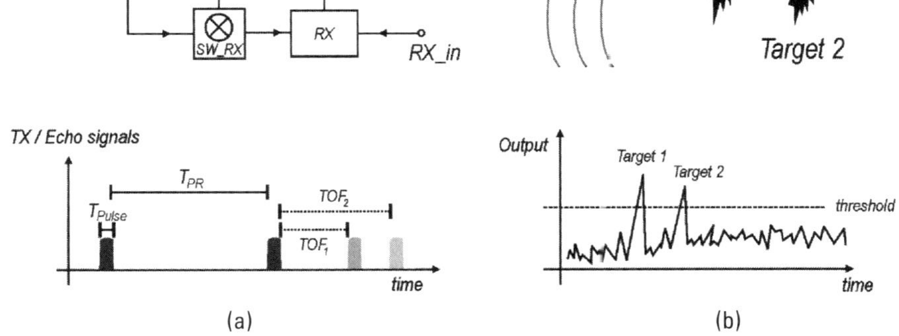

Figure 7.2 Operating principle of UWB radar sensor based on RX analog correlation: (a) transmitted/reflected pulses, and (b) output waveform for target detection.

at the RX output allows improving the sensor output level. The accumulation time, T_{INT}, and hence the number of integrations (n), are set according to the optimum threshold and probabilities of detection and false alarm (typically around 0.9 and 10^{-9}, respectively). The achievement of such probabilities highly depends on the SNR of the RX chain. Indeed, the noise level in the receiver may be orders of magnitude higher than the voltage level of the received pulse; thus, coherent collection of several hundred pulses may be needed before a decision threshold is reached. A detailed block diagram of the proposed SRR sensor is shown in Figure 7.3.

It consists of a RX/TX front-end and a customized local DSP (namely a control unit (CU)), which are integrated on the same IC. In the TX section, the carrier generation is implemented by means of a 24.125-GHz voltage-controlled oscillator (VCO) within a phase-locked loop (PLL). To improve the interference rejection and thus guarantee safe operation in multisensor networks,

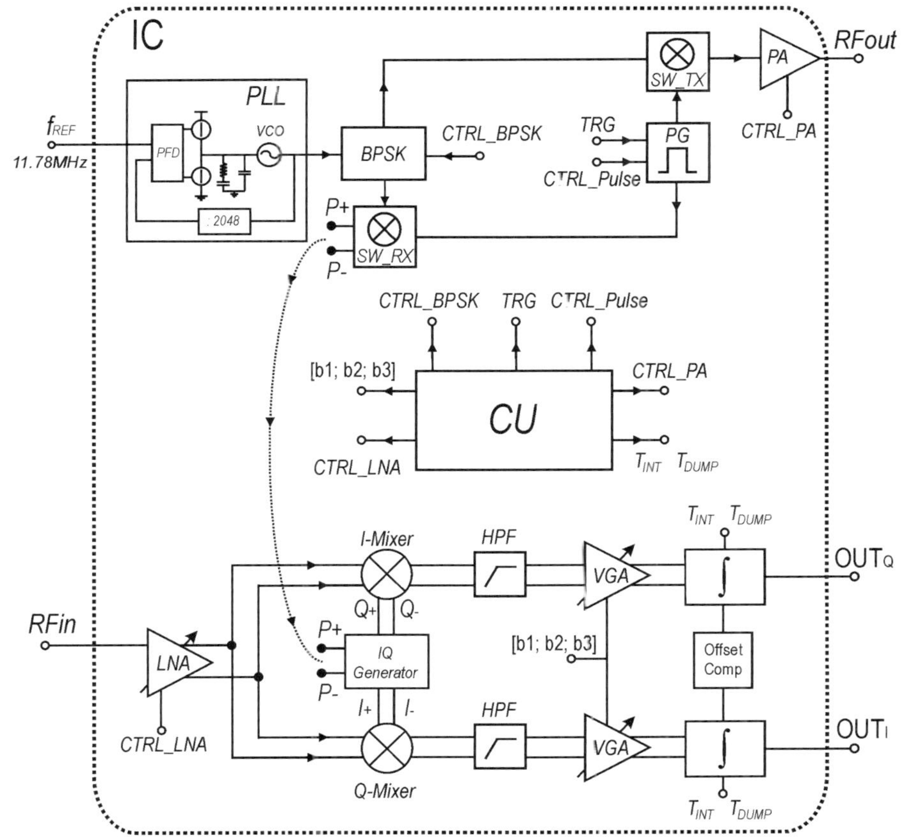

Figure 7.3 Block diagram of the UWB SRR sensor. (© 2009 IEEE. Reprinted with permission, from [7].)

the CW carrier is first BSPK modulated and then pulsed using switch SW_TX driven by pulse generator PG. The resulting 24-GHz UWB signal is properly amplified by the PA to comply with the emission mask regulations [29, 30]. The pulse generator produces the baseband rectangular pulses, whose T_{pulse}, and T_{PR} can be varied thanks to the digital control signal, CTRL_Pulse. Control signal, TRG, triggers the rectangular pulse, thus performing a time hopping (TH) modulation (i.e., pulse position modulation).

The RX performs a time correlation between the echo signal and a delayed replica of the TX pulse by means of the I/Q mixers, which are driven by switch SW_RX. The UWB correlation signal is high-pass filtered against interferences around the 24-GHz band (e.g., ISM, radio astronomy, and remote sensing applications) and then amplified by the variable-gain amplifier (VGA). Finally, it is accumulated into the offset-compensated I/Q integrators. The integration-and-dumping stage increases SNR by a factor n. After several integration periods, a reset signal discharges the integrator. For this system, the probability of target detection at a specific range increases as the number of collected pulses increases, provided that a particular time delay of the TX signal replica is maintained. At the design level, the RX trigger has to be kept in the same swept delay value for the number of needed accumulations. The in-phase (I) and quadrature (Q) outputs are then processed to extract the Doppler information (i.e., relative velocity), according to common radar signal processing methods discussed in Chapter 5.

The CU produces all the control signals required for pulse transmission and echo detection. Specifically, CU triggers the pulse generation and the BPSK modulation, controls the RX window timing, and sets the pulse width (T_{pulse}) and repetition (T_{PR}) as well as the RX gain. It also adjusts the integration (T_{INT}) and dumping (T_{DUMP}) times to integrate multiple signal replicas.

T_{pulse} and T_{PR} values are set according to FCC/ETSI regulations and SR requirements reported in Table 7.1.

The limitations on the equivalent isotropic radiated power density (EIRPD) impose a duty cycle δ, according to the following relationship

Table 7.1
Typical Automotive SR Requirements

Application	R_{res} [m]	R_{max} [m]	Relative Velocity [ms]
Parking aid	0.1–0.05	5	0–8
Blind spot	0.1	10	15–25
Stop and go	0.5	30	10
Precrash	0.1	10	50
Collision avoidance	0.2	30	50

$$\delta = \frac{T_{pulse}}{T_{PR}} = \frac{EIRPD_{avg}}{EIRPD_{peak}} = 0.37\% \tag{7.1}$$

where $EIRPD_{avg}$ and $EIRPD_{peak}$ are the maximum allowable values of average and peak EIRPD (i.e., 0 dBm/50 MHz and –41.3 dBm/MHz, respectively). Moreover, T_{pulse} and T_{PR} depend on the maximum unambiguous range, R_{max}, and target resolution, R_{res}, according to equations

$$R_{res} = \frac{cT_{pulse}}{2} \tag{7.2}$$

$$R_{max} = \frac{cT_{PR}}{2} \tag{7.3}$$

The most stringent values of R_{res} and R_{max} (i.e., a resolution better than 0.1m for parking aid and a maximum range of 30m for stop and go situations) correspond to a minimum T_{pulse} lower than 0.67 ns and a minimum T_{PR} of 200 ns, respectively. In this design T_{pulse} and T_{PR} can be set to 0.5 ns or 1 ns and 135 ns or 270 ns, respectively, by using control signal CTRL_Pulse produced by the CU.

7.2.2 PLL Circuit Design

In the proposed SRR sensor, the 24.125-GHz carrier is generated by a fully integrated PLL according to the simplified scheme shown in Figure 7.3. The PLL exploits a well-established architecture that is composed of a phase frequency detector (PFD), a charge pump (CP), a loop filter, a VCO, and an N-integer divider. An external reference at 11.78 MHz and a division ratio (N) of 2,048 are adopted to obtain the 24.125-GHz carrier. The PLL also generates a 3-GHz signal exploited in the pulse generation and the CU clock at 750 MHz.

The most critical blocks in the design of the PLL are the CP and the VCO. Indeed, the noise contributions and linearity performance of the CP strongly affect the PLL in-band phase noise and spur level, whereas the VCO determines the PLL out-band phase noise. The simplified schematic of the CP and VCO are shown in Figures 7.4 and 7.5, respectively.

The CP that is driven by the UP/DW signals generated by the PFD exploits a symmetric differential topology that allows minimizing the in-band noise contributions and spurs of the PLL. Cascode MOS mirrors are included

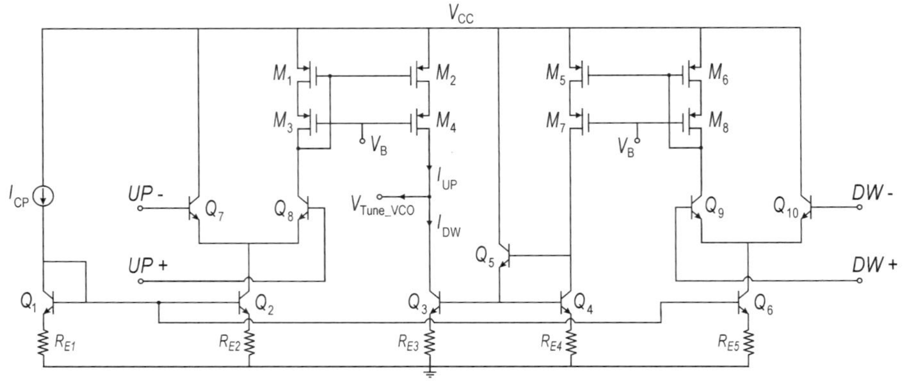

Figure 7.4 Simplified schematic of the CP.

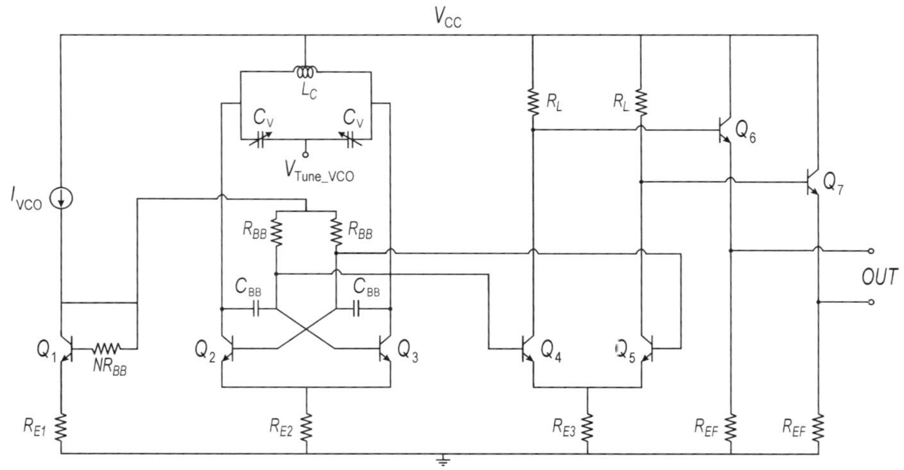

Figure 7.5 Simplified schematic of the 24-GHz VCO. (© 2008 IEEE. Reprinted with permission from [31].)

to improve the current matching and enhance the output impedance. As demonstrated in [32], the overall in-band PLL phase noise due to MOS current mirrors, emitter resistances, and bipolar transistors reduces by increasing bias current I_{CP}. At this purpose, I_{CP} is set to 200 μA. The CP must also guarantee the output swing required to the VCO to maintain the frequency locked at 24.125 GHz despite process tolerances, which impact the LC resonator.

The VCO adopts a differential cross-coupled architecture. The oscillation frequency is set by the parallel LC resonator, which adopts A-MOS varactors and a single-turn multilayer inductor, L_C. At these frequencies, A-MOS varactors exhibit good Q-factor values and are preferable to diode varactors due to the higher tuning capability and simplicity of connection. Both the cross-coupled pair and the bias scheme contribute to the VCO phase noise. The size of active devices should properly be optimized by minimizing the sensitivity function of the frequency with respect to the bias current. A proper bias circuit is arranged to reduce the active-stage noise contributions. Indeed, the VCO core is part of a current mirror with degeneration resistances, thus removing the need for current tail sources that are noisy. Actually, emitter degeneration resistors at 24 GHz instead of current sources provide both wider oscillation amplitude and comparable or even higher common mode rejection. A buffer made up of a differential stage with resistive load is used to connect the oscillator core with differential emitter followers. This allows the avoidance of load effects on the VCO and provides driving capability for the following blocks (i.e., prescaler and BPSK modulator).

The PLL division ratio N (i.e., 2,048) is implemented by means of 11 divide-by-two stages. The most common architectures for a high-speed frequency divider include static frequency divider, Miller frequency divider, injection-locked frequency divider, and regenerative structures. In this design, a static frequency divider is preferred due to its wider frequency range and better robustness with respect to the other architectures. Each divider stage consists of ECL flip-flops (i.e., D-latches) in a closed-loop master-slave configuration and emitter-follower stages, as shown in Figure 7.6. Due to the large parasitic capacitance, especially for the first divider stages, accurate post-layout parasitic extraction has to be performed to guarantee reliability at 24 GHz. Design parameters such as bias currents and load resistances of high-frequency divider stages are included in Figure 7.6.

Finally, the loop filter is designed to give an acceptable compromise between the PLL bandwidth and noise. Specifically, a 200-kHz bandwidth is set in this implementation. The filter adopts a dominant-pole capacitor of 480 pF with a series resistor of 6.4 kΩ for the zero, and a parallel capacitor of 30 pF for the second pole. The 480-pF capacitance is implemented using both MOM and MIM capacitors, whereas the other one is a single MIM device.

The measured output spectrum of the PLL is shown in Figure 7.7. It is centered at 24.125 GHz and presents two spurious tones at a frequency offset of 11.78 MHz (i.e., the PLL reference frequency, f_{REF}), whose levels are 50 dB lower. The VCO exhibits a tuning range of 4.7 GHz from 20.4 to 25.1 GHz. The PLL reports the same performance when f_{REF} sweeps between 9.78 and 12.2 MHz. The measured phase noise is –104.3 dBc/Hz at 1-MHz offset from the carrier at 24.125 GHz.

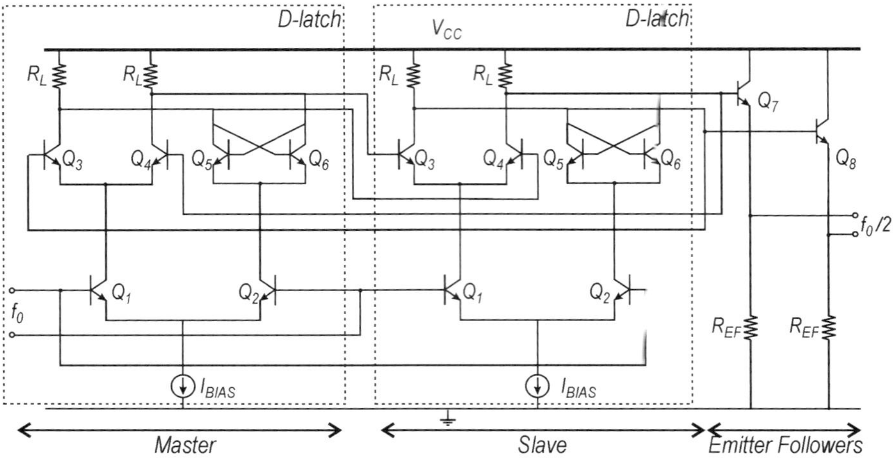

Divider stage	f_0 [GHz]	I_{BIAS} [mA]	R_L [Ohm]
1	24	1	200
2	12	0.5	400
3	6	0.33	600

Figure 7.6 Schematic of divider-by-two stage and design parameters of high-frequency divider stages (Source: [9]. Reprinted with kind permission from Springer Science and Business Media).

7.2.3 RX Circuit Design

As previously shown in Figure 7.3, the RX section of the UWB radar sensor consists of a variable-gain LNA followed by I/Q direct-conversion mixers, which are connected by means of high-pass filters to the IF VGAs and offset-compensated integrator and dump stages. The LO port of mixers are driven by the I/Q generator with a rectangular pulse train generated by the TX_SW. The simplified schematic of the down-converter (i.e., LNA and I/Q mixers), along with the I/Q generator, is shown in Figure 7.8.

The circuit exploits a fully differential topology to avoid the detrimental effects of parasitic ground and power supply inductances, while providing high rejection to common-mode spurious signals and substrate noise. Single-ended-to-differential conversion is provided by the on-chip RF input balun T_{in}. The three-stage LNA is coupled by an integrated transformer to a Gilbert quad pair using resistive loads (R_L = 400 Ω). Each amplifier stage adopts a transformer-loaded cascode topology to guarantee both high stable gain and reverse isolation. Simultaneous minimum noise and 50-Ω input impedance matching is obtained

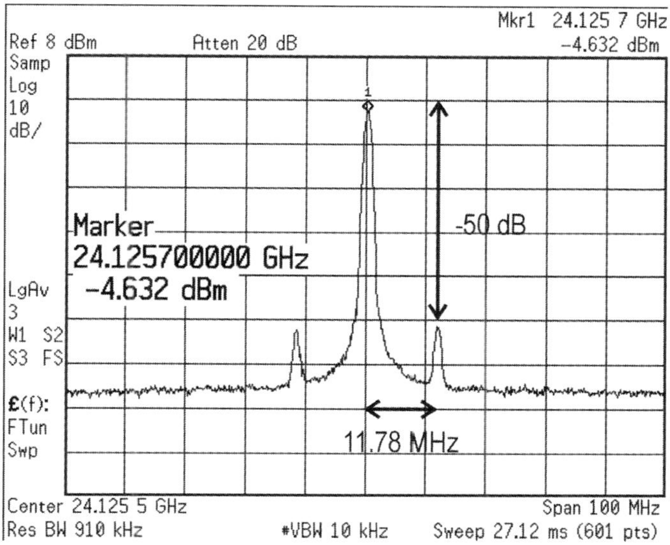

Figure 7.7 Measured output spectrum of the PLL. (© 2008 IEEE. Reprinted with permission from [31].)

by properly sizing the bipolar input pair and inductors L_{E1} and capacitor C_{in} [33], respectively. In order to trade off both gain and linearity performance, the second and third stage of the LNA adopt inductive emitter degeneration, L_{E2} and L_{E3}. Transformers T_{L1}, T_{L2}, T_{L3}, together with shunt MIM capacitors C_{P1}, C_{P2}, C_{P3}, provide the maximization of both gain and power transfer between stages, while preserving the required bandwidth flatness. To avoid RX saturation in case of high input power levels, gain-variation functionality (i.e., 20-dB gain variation) is implemented in the second and third stages by means of an effective current-steering topology. Indeed, the control voltage V_{CTRL_LNA} sets low and high gain levels, respectively. This approach significantly increases the receiver linearity, still maintaining excellent noise figure performance. To further optimize the linearity performance of the whole down-converter, the LNA is transformer-coupled to the Gilbert quad-pair that is directly driven by the secondary coil signal currents of T_{L3}, thus avoiding the adoption of traditional voltage-to-current converters. It is worth noting that this circuit arrangement is largely exploited also in 77-GHz radar RXs, which highly suffers of linearity issues (see Section 7.5). Down-converter bias currents I_{LNA1}, I_{LNA2}, I_{LNA3}, and I_{MIX} are 13, 15, 18, and 14 mA, respectively.

The I/Q generator takes advantage of a second-order RC polyphase filter [34], whose time constant is properly sized for a 24.125-GHz operating frequency. The filter uses poly resistors R_F of 66 Ω and MIM capacitors C_F as low as 100 fF. Amplifiers (G_I, G_Q) are included to compensate for the loss due to

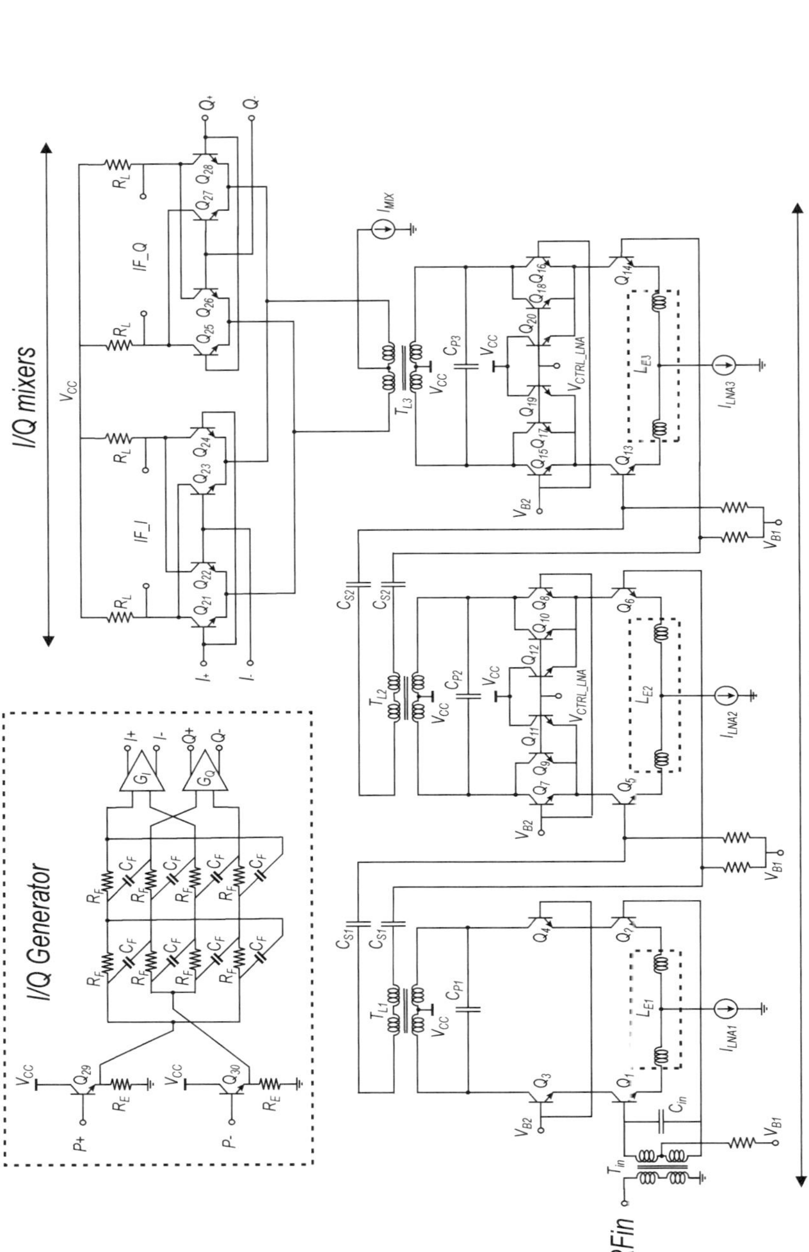

Figure 7.8 Simplified schematic of the down-converter with the I/Q generator. (© 2009 IEEE. Reprinted, with permission, from [7].)

both the polyphase filter (about 3 dB) and the LO paths (about 0.5 dB) toward the Gilbert quads. To minimize the phase error between I and Q signals, the LO paths are laid out in close proximity and take advantage of an interlaced structure [9].

To mitigate the effects of FMCW radar interference at 24.125 GHz, a high-pass filtering is required before the IF amplification stages. To this aim, I/Q mixer outputs are ac-coupled to the IF VGAs. By setting the coupling capacitor to 1 pF, a cut-off frequency of around 330 MHz is achieved that is a trade-off between interferer/signal losses.

The down-converter achieves a conversion gain as high as 30 dB and a 6-dB noise figure, as shown in Figure 7.9. The gain drop around 24.125 GHz is due to the adopted ac-coupling at the mixer outputs.

Thanks to the LNA gain-control functionality, the down-converter achieves an input 1-dB compression point (IP1dB) of –15 dBm. At low-gain setting, it exhibits a 10-dB power gain, while the noise figure increases up to 10.5 dB.

The design of the baseband section is also crucial for the radar sensor implementation since it has to guarantee the largest part of the variable gain required by the application. To this aim, a two-stage VGA can be employed in both I/Q paths. A solution is shown in Figure 7.10. Each VGA consists of two differential cascode stages with digitally selectable load resistances. By means of control bits (b1 to b3), a dB-linear gain characteristic is achieved with around 30 dB of gain variation (i.e., 6 to 36 dB) in a 2-GHz bandwidth. The switches are implemented by n-MOS transistors, whose widths are set for the best tradeoff between R_{ON} reduction and the fulfillment of the demanding bandwidth requirements. Degeneration resistor R_{EE} of 200 Ω and a biasing current I_{VGA} of 3.8 mA are set to trade off linearity performance and power consumption. To

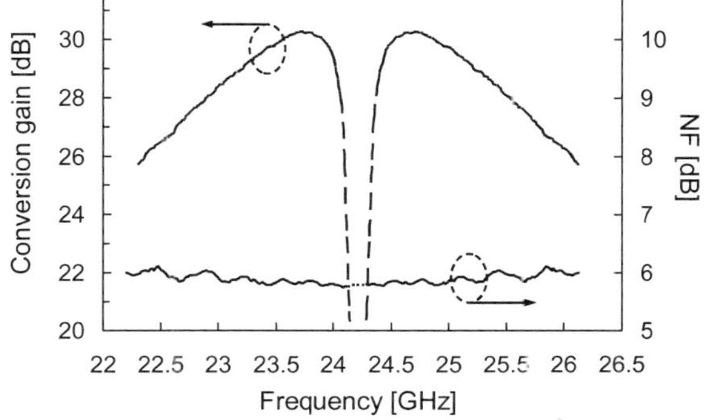

Figure 7.9 Down-converter power gain and noise figure at high-gain setting. (Source: [9]. Reprinted with kind permission from Springer Science and Business Media.)

ensure a stable biasing point at the output, a common mode feedback circuit (CMFB) with resistive sensing is also included.

The RX section is completed by two Gm-C integrate-and-dump stages that perform a coherent collection of the received pulses, thus producing the output detection signal at I/Q terminals. This properly increases SNR and improves the detection probability with a low false-alarm factor. The integration is repeated n times while preserving the stored charge for the whole T_{PR} without exhibiting any significant discharge phenomena. Due to the relatively high gain of the baseband amplifiers, the RX chain could suffer from dc offsets that can highly degrade the echo detection. Therefore, an offset compensation circuit is required to reduce the false alarm probability due to erroneous overcoming of the detection threshold.

Figure 7.11 shows the integrator with the proposed offset-compensation circuit.

As in the VGA, a CMFB circuit guarantees a stable bias point (C_1 and C_2 are compensation capacitors). The input voltage pulses are integrated according to the well-known law

$$V_{out} = \frac{g_{m1,2}}{C} \int V_{in} dt \tag{7.4}$$

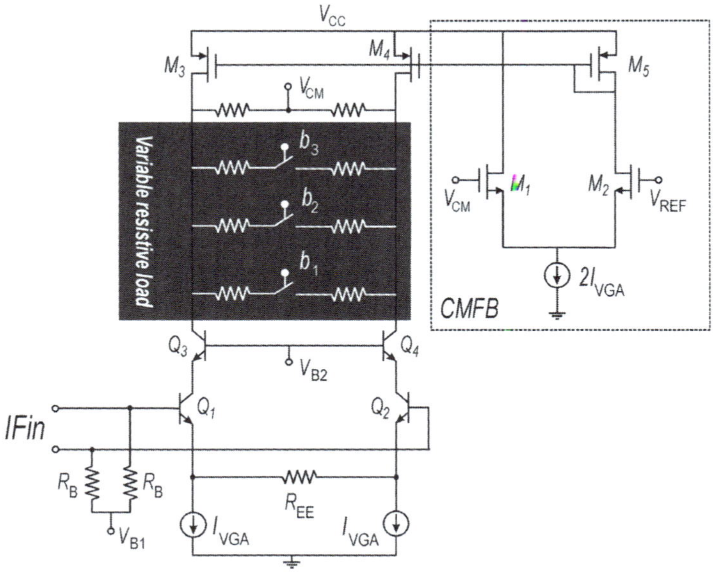

Figure 7.10 Simplified schematic of a gain stage of the VGA. (© 2007 IEEE. Reprinted with permission from [35].)

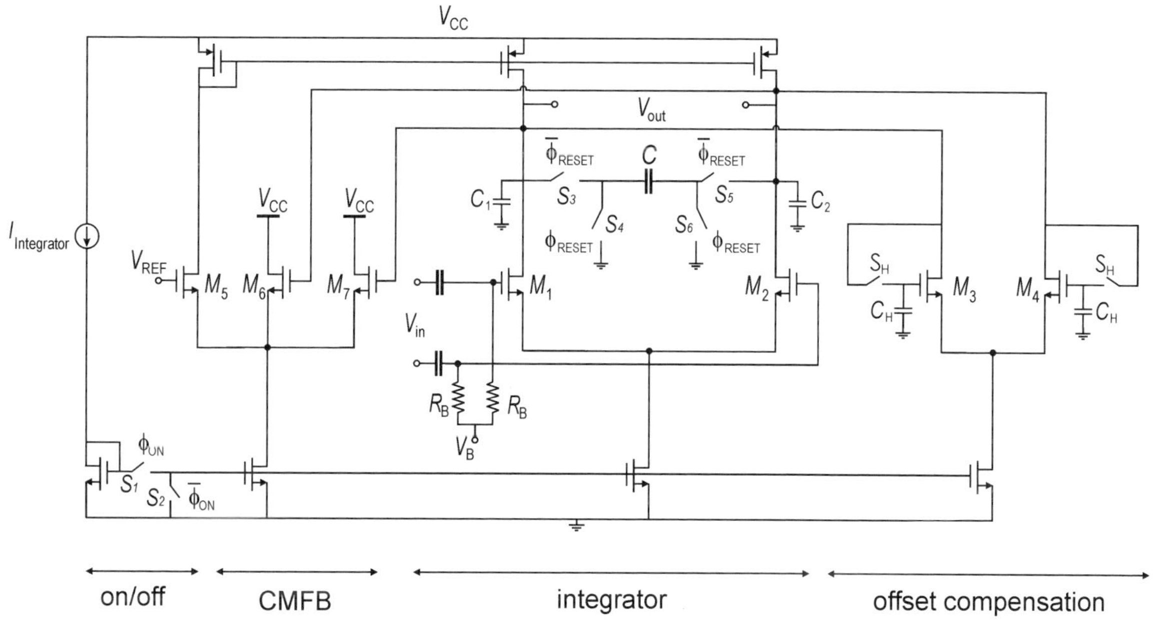

Figure 7.11 Simplified schematic of the integrator. (© 2007 IEEE. Reprinted with permission from [35].)

where $g_{m1,2}$ is the transconductance of transistors $M_{1\text{-}2}$ and C is the integrator capacitance (i.e., an MIM capacitor of about 3 pF). The on/off switches S_1 and S_2 are used to avoid discharge of capacitor C during the TPR. Control signals of S_1 and S_2 (i.e., $\Phi_{ON}, \overline{\Phi}_{ON}$) are synchronous with the signal coming from the PG (through the SW_RX) toward mixer LO ports. On/off switches are necessary because the system works with an extremely low duty cycle (i.e., 0.37%) and then the integrator has to maintain the charge accumulated during T_{pulse} for the whole T_{PR}. After several integration periods, the reset signal, RESET, discharges the capacitor for a T_{DUMP} time and a new cycle of integration can start. Transistor widths of the reset switches S_3, S_4, S_5, and S_6 should properly be reduced to limit the charge injection phenomena into the output capacitor. Both the integration and dump times (i.e., T_{INT} and T_{DUMP}) are properly set by the CU that also enables the RESET signal.

As far as the offset compensation is concerned, the technique in Figure 7.11 allows reducing the input-referred offset according to the following equation:

$$V_{OS\_IN} \approx \frac{V_{OS1,2}}{g_{m3,4} r_o} + \frac{V_{OS3,4}}{g_{m1,2} r_o} \tag{7.5}$$

where $V_{OS1,2}$ ($V_{OS3,4}$) and $g_{m1,2}$ ($g_{m3,4}$) are the offset and the transconductance of differential pair $M_{1,2}$ ($M_{3,4}$), respectively, and r_o is the integrator output resistance [35].

The output voltage of the integrators is a function of the overall RX gain, the Gm/C ratio, and the integration time T_{INT}. Figure 7.12 shows the measurement of the integrator output voltage at high gain setting, as a function of different VGA gain levels (i.e., six levels from 36 dB to 6 dB) for a 1-ns pulse. The input pulse energy (E_{pulse}) is –143 dBmJ. T_{PR} is set to 270 ns. T_{INT} of 885 μs and a reset signal of 10 μs are adopted. The behavior of the integrator in a single integration period is visible in the inset of Figure 7.12, which demonstrates that the integrator is able to preserve the stored charge up to the next integration. Of course, the sensor output voltage is directly proportional to the correlation percentage between the on-chip replica (at the LO port) and TX pulses, as shown in Figure 7.13. The sensor achieves a maximum output voltage of 280 mV when pulses are completely overlapped.

7.2.4 TX Circuit Design

The TX is designed to produce a 24-GHz modulated pulse train in compliance with both FCC (United States) and ETSI (EU) spectrum masks. Additionally, TX EIRP peak in a 50-MHz band centered at 24.125 GHz should be below 0 dBm.

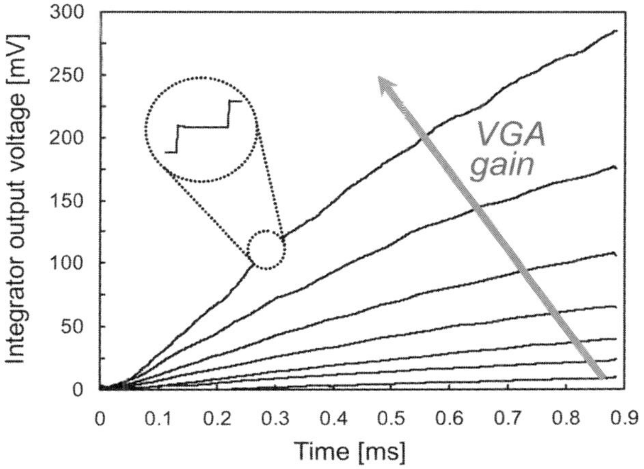

Figure 7.12 Measured output voltage of integrators (T_{pulse} = 1 ns, E_{pulse}= −143 dBmJ, T_{PR} = 270 ns, T_{INT} = 885 μs, T_{DUMP} =10 μs, TX/RX correlation = 100%). (© 2009 IEEE. Reprinted with permission from [7].)

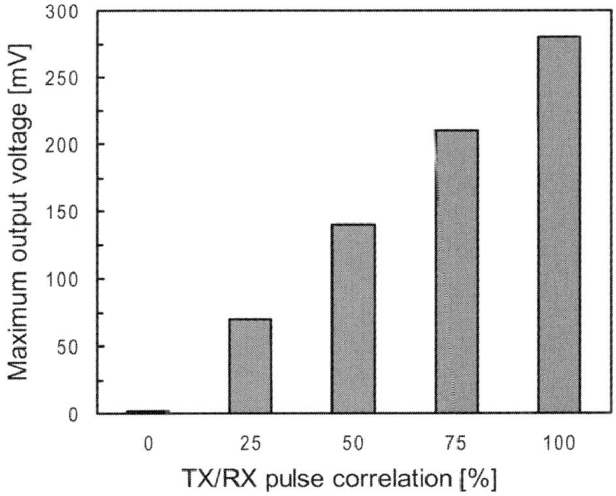

Figure 7.13 Measured output voltage of integrators at different RX/TX pulse correlation percentage (T_{pulse} = 1 ns, E_{pulse}=-143 dBmJ, T_{PR} = 270 ns, T_{INT} = 885 μs, T_{DUMP} =10 μs).

A crucial block in pulsed radars is the pulse generator, since it determines the specifications of the entire sensor in terms of R_{max} and R_{res}, according to (7.2) and (7.3). The adopted solution for the PG allows the baseband pulse train to be provided with two different pulse widths (i.e., 1 ns and 0.5 ns, corresponding to R_{res} of 15 cm and 7.5 cm and T_{PR} of 135 ns and 270 ns, respectively) that limit the maximum unambiguous range to 20m and 40m, respectively. These settings fulfill typical requirements of SRR application, previously reported in Table 7.1. The architecture of PG is shown in Figure 7.14.

The baseband pulse is produced by means of a NAND operation between the TRG signal properly delayed (A/B) and the inverted TRG signal (C). All adopted latches use the 3-GHz clock coming from the PLL. Each latch produces a delay of 333 ps or 166 ps if the phase shift between its clock (Cp , Cn) and the one of the retimer is 0 or 180 degrees, respectively. Two different delay paths are available (i.e., 999 ns or 499 ns), which are selected by means of the CTRL_pulse of the multiplexer (MUX).

The simplified schematics of the TX blocks (i.e., the BPSK modulator, the SW, and the PA) are depicted in Figure 7.15.

The BPSK modulator is based on a differential pair, Q_1-Q_2, driven by the RF signal and a current-steering switch, Q_3-Q_6, driven by the CTRL_BPSK signal. The modulated RF signal drives the current-steering switch SW_TX that produces the 24-GHz modulated UWB pulse. Finally, the PA boosts the

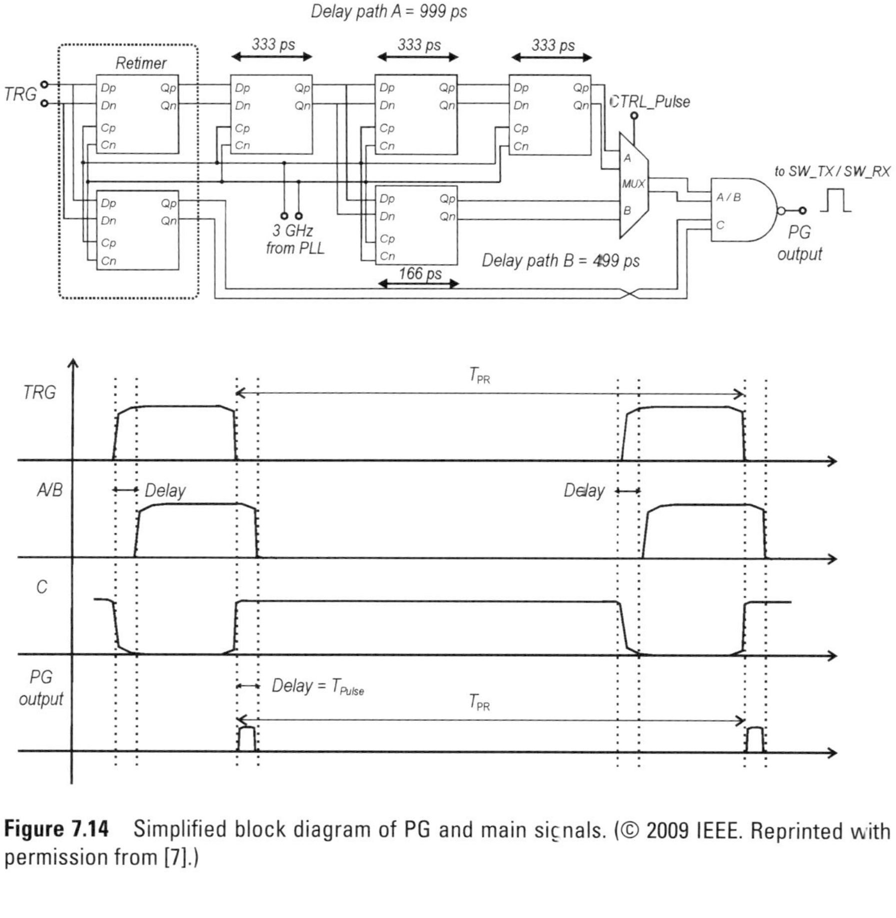

Figure 7.14 Simplified block diagram of PG and main signals. (© 2009 IEEE. Reprinted with permission from [7].)

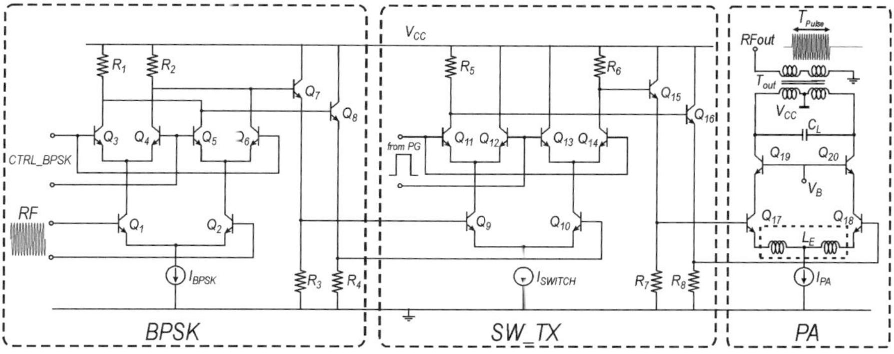

Figure 7.15 Simplified schematics the TX. (© 2009 IEEE. Reprinted with permission from [7].)

output signal level to be compliant with the regulation limits [29, 30]. To this aim, an antenna gain of 10 dB is considered. The PA adopts a differential cascode topology with transformer resonant load. The bias voltage of the common-base transistors is designed to provide a low base impedance (of about 150 Ω), which boosts the effective breakdown voltage from 1.8V (BV_{CEO}) to 3.1V (BV_{CER}). The PA delivers 0 dBm or 3 dBm output power by setting the bias current I_{PA} for T_{pulse} of 1 ns and 0.5 ns, respectively. Transformer T_{out} provides differential-to-single-ended conversion of the output signal. Moreover, thanks to its inherent galvanic isolation, T_{out} also guarantees robustness, thus avoiding ESD protection structures that greatly affect the RF performance.

Figure 7.16 shows the measured power spectral density (PSD) and the simulated time-domain waveforms at the TX output considering an antenna gain of 10 dB for both T_{pulse} of 1 ns and 0.5 ns. The T_{PR} is set to 270 ns and 135 ns, respectively. The spectra present the main lobe is centered at 24.125 GHz and two nulls, whose frequency span is 2 GHz and 4 GHz for 1 ns and 0.5 ns pulses, respectively. The PSD curves are compliant with the maximum allowed EIRP defined by the ETSI mask [30].

7.2.5 On-Chip Inductive Component Design

Low-loss inductive components are of paramount importance for a successful implementation of millimeter-wave ICs [6]. Of course, the availability of an optimized BEOL is crucial since top copper metals and thick intermetal oxides are mandatory to lower ohmic losses in the metals and reduce parasitic capacitances toward the substrate, respectively. On the other hand, ad hoc design of both inductors and transformers instead of using library components is still a plus when performance has to be considerably pushed.

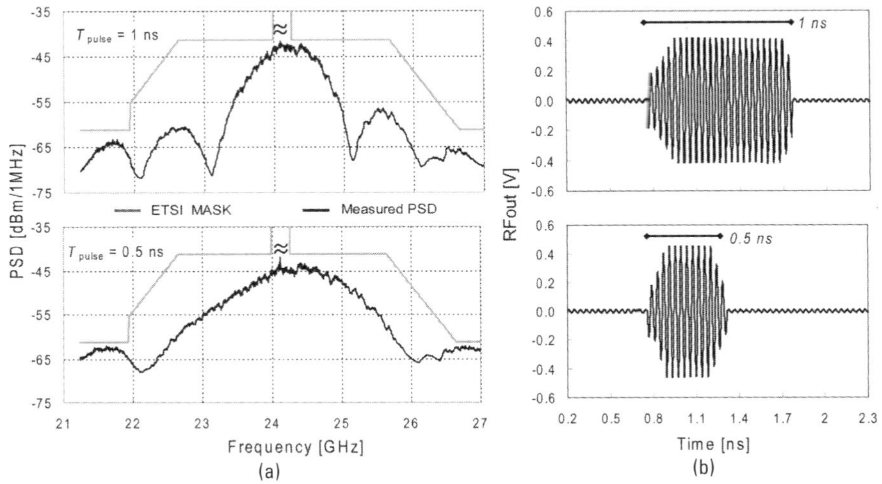

Figure 7.16 (a) Measured PSD and (b) simulated time-domain waveforms of transmitted pulses. (© 2009 IEEE. Reprinted with permission from [7].)

Three-dimensional (3D) views of inductive components used in the radar sensor, along with the BiCMOS technology metal stack, are shown in Figure 7.17. Since an extensive exploitation of transformers is done in both TX and RX sections, it is useful to summarize geometrical and electrical features of the adopted transformers, as shown in Table 7.2.

In the K-/Ka-bands (i.e., up to 40 GHz), the adoption of a polysilicon patterned ground shield (PGS) beneath spirals still has considerable effects in the reduction of substrate losses [36, 37], without significantly affecting the

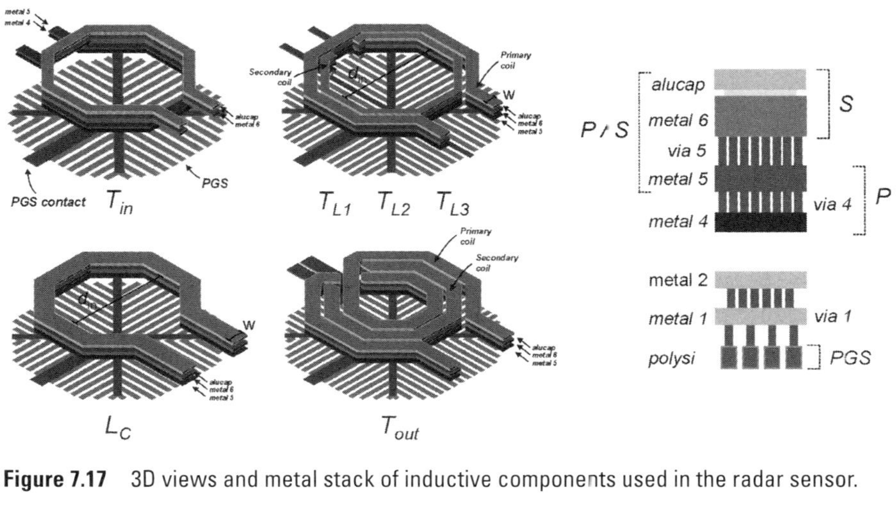

Figure 7.17 3D views and metal stack of inductive components used in the radar sensor.

Table 7.2
Transformer Geometry and Electrical Data at 24 GHz

	T_{IN}		T_{L1}, T_{L2}, T_{L3}		T_{OUT}	
Transformer	**Primary (P)**	**Secondary (S)**	**Primary (P)**	**Secondary (S)**	**Primary (P)**	**Secondary (S)**
Metal layers	M4-M5	Al-M6	Al-M6-M5	Al-M6-M5	Al-M6-M5	Al-M6-M5
w [μm]	5	5	5	8	10.5	10.5
L [nH]	0.25	0.27	0.34	0.26	0.46	0.19
d_{in} [μm]	65		73		43	
k	0.7		0.4		0.5	
Structure	Stacked		Interleaved		Interleaved 2:1	

operative frequency range (i.e., reduction of the self-resonance frequency (f_{SR})). In our process, the PGS is built using the available doped polysilicon layer under the metal 1, which is used for shield grounding. The PGS is profitably included in both inductors and transformers. Series losses are instead reduced by means of proper exploitation of multilayer spirals, thanks to the availability of six metallization layers. As far as the geometrical parameters are concerned, single-turn spirals and small inner diameters (d_{in}) are generally used, since at these frequencies sub-nH inductors are required [38]. Narrow metal width (w) is also preferred to reduce the capacitance toward the substrate, provided that electromigration limitations are fulfilled.

As far as integrated transformers are concerned, most suitable structures to optimize main design parameters are the traditional interleaved and stacked configurations. At millimeter-wave frequencies, interleaved configurations can be profitably used to implement symmetric windings with coupling factor (k) as high as 0.5 by taking advantage of the top metals available in the process. In such a way, both f_{SR} and Q-factor are maximized at the expense of the area consumption. On the other hand, stacked transformers achieve higher k and a better area exploitation, but exhibit lower f_{SR} due to the increased parasitic capacitances of bottom metals. The most important performance parameter for transformers in circuit applications is the capability to transfer power from the input to the output port with low loss. This power transfer is related not only to the transformer itself, but also to the impedances connected to its input and output, including matching networks. In the following, a more detailed description of the inductive components depicted in Figure 7.17 is provided to the reader.

In the RX section, transformer T_{IN} is used for single-to-differential conversion and 50-Ω impedance matching. The reference design parameter for T_{IN} is the insertion loss (IL) that is highly related to k, and it is affected by coil losses. Therefore, T_{IN} takes advantage of a stacked configuration to maximize

k, while both primary and secondary windings exploit a two-layer structure to reduce the coil series resistance.

On the other hand, each LNA stage uses an interleaved transformer (i.e., $T_{L1\text{-}3}$) whose coils exploit a three-layer structure consisting of two highest Cu metals (metal 6 and metal 5) plus a top alucap layer. The design of $T_{L1\text{-}3}$ is based on the maximization of the transformer characteristic resistance (TCR) [39] as it is related to the achievable power gain of the LNA [31]. In a first approximation the TCR is given by

$$TCR = \omega\ Q_{EQ} L_{EQ} = \omega \cdot \left(Q_P \frac{k^2\ Q_P\ Q_S}{1 + k^2\ Q_P\ Q_S} \right) \cdot \left(L_P + \frac{L_P}{Q_S^2} + k^2 \frac{L_P \cdot Q_S}{Q_P} \right) \quad (7.6)$$

where L_P, Q_P and L_S, Q_S are the inductance and the Q-factors of primary and secondary coils, respectively. The use of a three-layer interleaved 1:1 configuration allows achieving a TCR higher than 900 Ω at 24 GHz.

In the TX section, an interleaved configuration with 2:1 turn ratio is used for transformer T_{out} with the aim of attaining power matching and differential to single-ended conversion at the TX output. The adopted structure also allows improving the magnetic coupling that results higher than the one of $T_{L1\text{-}3}$. Moreover, the use of a multilayer structure for both coils, along with a slightly higher metal width, allows reducing the coil series resistances.

Finally, the LC tank of the VCO adopts a single-turn inductor, L_C, whose value at 24 GHz is as low as 230 pH. It uses the same three-layer structure of interleaved transformers and achieves a Q-factor as high as 19 at 24 GHz, which is essential for the VCO phase-noise performance.

It is worth mentioning that geometrical parameters of inductive components are first defined by using scalable lumped models [40] that allow the designer to maintain a close contact to the physical effects. Then model validation is carried out by means of EM simulations, taking into account the coupling effects due surrounding structures (i.e., ground plane, connections, and so on).

7.2.6 Radar Sensor Implementation

The micrograph of the overall radar sensor IC is shown in Figure 7.18. The chip area is pad limited and the die size is 3 mm × 3 mm. Main circuit blocks discussed in the previous sections are highlighted in the picture.

The sensor has been conceived for a flip-chip assembly on a commercial dialectic substrate with ε_r of 2.33 (i.e., ARLON 25FR). The same substrate is also used to design the RX/TX antennas that exploit an aperture-coupled patch array structure. Figure 7.19 depicts the 3D view and cross-section of the assembled radar sensor, along with the detail of the patch-array antenna geometry

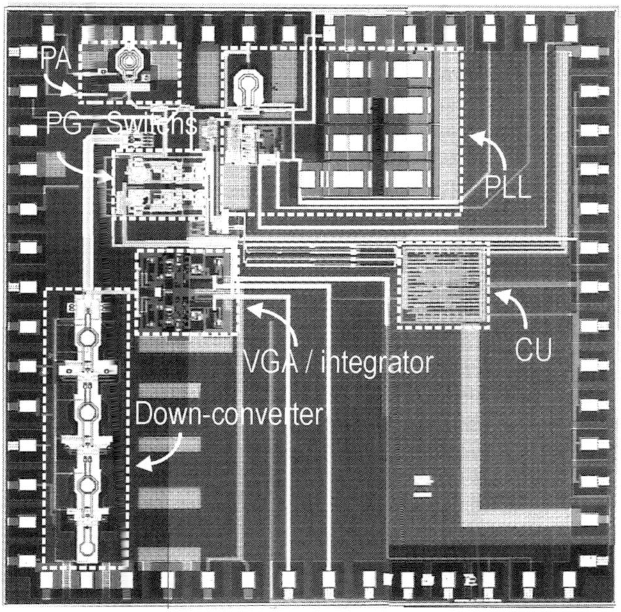

Figure 7.18 Radar sensor IC micrograph. (© 2009 IEEE. Reprinted with permission from [7].)

and its radiation pattern. The stacked structure allows very low sidelobes to be obtained, while the feed network is completely isolated from the radiating elements. Moreover, the aperture coupled patch-antenna configuration guarantees a broad frequency bandwidth. The number of patch elements and the current distribution should be set to comply with the ETSI radiation mask. As shown in the radiation diagram in Figure 7.19(d), a maximum gain of about 10 dB is achieved with a sidelobe level of –30 dB. The frequency bandwidth is better than 2.4 GHz. Finally, the radar sensor performance is summarized in Table 7.3.

7.3 Transmitter Chipset for 24-/77-GHz Automotive Radar Sensors

For a long time ADASs addressed a very limited market, mainly focused on high-end cars. Indeed, the pervasive adoption of radar technology has to comply with a challenging trade-off between performance and cost, which is essential to engage the mass market of low-end cars. From this perspective, the adoption of dual-band millimeter-wave transceivers, implemented in a well-established silicon-based technology, can improve the cost-effectiveness of ADASs.

This section is dedicated to the description of a dual-band TX chipset for 24-/77-GHz automotive radar. The proposed solution enables the LR operation

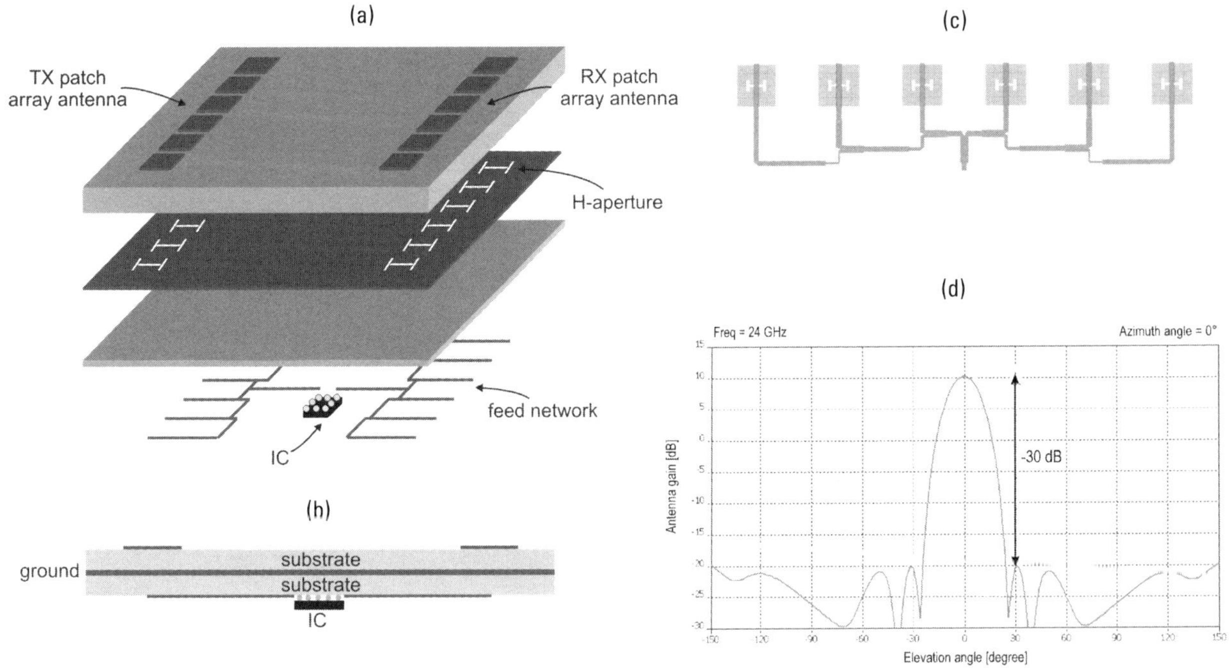

Figure 7.19 Sensor assembly: (a) 3D view, (b) cross-section, (c) antenna geometry, and (d) antenna radiation pattern.

Table 7.3
Summarized Performance of the Radar Sensor

RF frequency	22 to 26 GHz
IF frequency	0.1 to 2 GHz
PLL reference frequency	11.78 MHz
RX input return loss (S_{11})	–20 dB
Down-converter gain	30 / 10 dB
Down-converter noise figure	6 dB
Down-converter IIP_{1dB}	–15 dBm
VGA gain	6 to 36 dB
VCO phase noise at 1-MHz offset	–104 dBc/Hz
VCO tuning range	20.4 to 25.1 GHz
PLL spur level	–50 dBc
PA power at 24 GHz	3 / 0 dBm
Sensor output voltage	8 to 280 mV (E_{pulse}=–143 dBmJ)
UWB bandwidth	1.3 GHz (T_{pulse}=1 ns, T_{PR}=270 ns) 2.1 GHz (T_{pulse}=0.5 ns, T_{PR}=135 ns)
Current consumption (Vcc = 2.5 V)	32 mA (frequency synthesizer) 25 mA (BPSK+PG+switches) 7 mA (PA) 65 mA (I/Q down-converter) 19 mA (VGAs +Integrators)
Silicon area	9 mm^2
Max antenna gain at 24 GHz	10 dB
Antenna frequency bandwidth	2.4 GHz

at 77 GHz in a standard SiGe HBT BiCMOS technology (i.e., BiCMOS9 [1]) by taking advantage of building blocks developed for the 24-GHz SRR sensor [7, 8]. Although the adopted process is not suitable for W-band applications in terms of f_T/f_{max} and BEOL, ad hoc arrangements (i.e., proper current density for the PA devices, ground plane parasitic cancellation, and optimized transformer design) allow state-of-the-art millimeter-wave PA performance to be achieved, thus demonstrating low-cost dual-band radar TX.

Figure 7.20 shows the block diagram of a 24-/77-GHz TX chipset. The SRR TX generates an UWB signal by means of the 24-GHz PLL previously described in Section 7.2.2. The sub-ns switch is designed to produce a 24-GHz modulated pulse train with proper T_{PR} and T_{pulse} to cover main sensor requirements for SRR applications, as detailed in Section 7.2.1. In the proposed architecture, the SRR TX is also used to drive the LRR TX by using the 25.5-GHz CW signal at the PA output. The LRR TX takes advantage of the millimeter-wave mixer, which up-converts the IF signal from 25.5 GHz to 76.5 GHz using the 51-GHz LO signal produced by a frequency doubler. The adopted

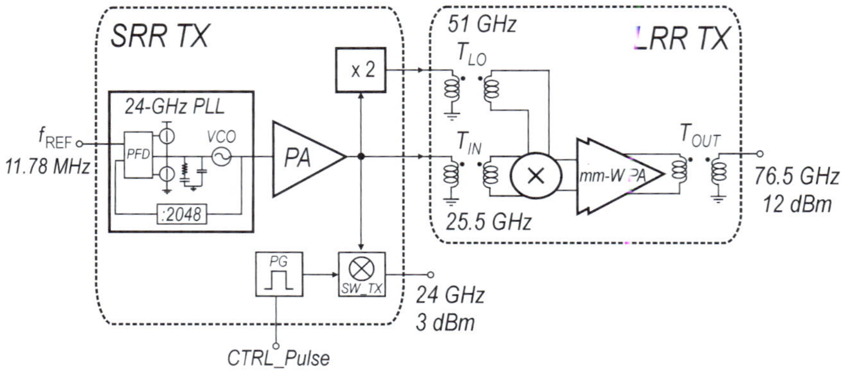

Figure 7.20 Block diagram of the dual-band 24-/77-GHz TX chipset for automotive radar sensors. (© 2010 IEEE. Reprinted with permission from [12].)

architecture and the selected frequency plan allow sharing the same PLL of the SRR section to produce both IF (directly) and LO (multiplying by two) signals. Finally, a two-stage millimeter-wave PA delivers the required output power level. On-chip transformers (T_{IN}, T_{LO}, and T_{OUT}) are used at circuit block interfaces to provide electrostatic discharge (ESD) protection, impedance/power matching, and single-ended-to-differential conversion.

7.3.1 Design of the 77-GHz TX Front-End

Figure 7.21 depicts the circuit schematic and the transformer structures of the LRR TX front-end. The up-conversion mixer is designed to properly drive the millimeter-wave PA in terms of both input impedance and power level. It adopts a fully differential topology consisting of a V-I converter (Q_1-Q_2) and a double-balanced Gilbert quad (Q_3-Q_6). Degeneration inductors L_E (implemented with folded microstrips) are used to improve mixer linearity and adjust its input impedance. A resonant load consisting of transformer, T_1, and capacitors, C_{SM}, is adopted to achieve the required conversion gain. Impedance matching between the up-converter and the PA is also guaranteed. Finally, at IF and LO ports center-tapped transformers, T_{IN} and T_{LO}, implement single-ended-to-differential conversion and 50-Ω conjugate matching. The millimeter-wave PA is made up of a driver and a power stage. It adopts a transformer-based pseudo-differential configuration that highly reduces the effect of parasitic inductances of power supply and ground, while avoiding the tail current voltage drop of a true differential structure. Each PA stage is based on a differential cascode (CAS) topology, which guarantees a higher stable gain, better reverse isolation, and enhanced robustness compared to the common-emitter (CE) topology. In particular, in a CAS topology, the use of the common-base (CB) in series to the CE increases breakdown voltage, thus improving the output power despite the

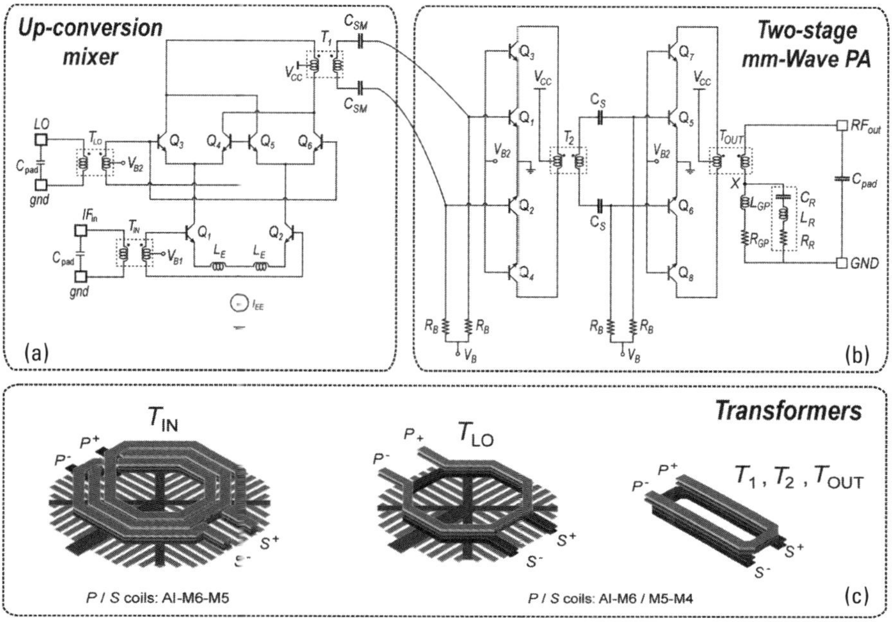

Figure 7.21 Schematic of the LRR TX front-end and 3D views of on-chip transformers. (Source: [27]. Reprinted with permission.)

additional voltage drop. Transformer T_2, together with series capacitors C_{SM}, provide optimum load for the first stage for better gain, whereas T_{OUT} maximizes the power delivered to the output load.

Given the relative low f_T/f_{MAX} (i.e., 166/175 GHz) of the adopted technology, both driver (Q_1–Q_4) and power-stage transistors (Q_5–Q_8) are biased at the peak-f_T current density (i.e., around 7 mA/μm^2 in this process) to maximize both gain and output power. This approach differs from the traditional design procedure for millimeter-wave PAs [15, 41], which suggests setting the bias current density at the peak f_T/f_{MAX} only for the driver stage (to maximize the small-signal power gain), while using a reduced current density (i.e., 50% of peak f_T/f_{MAX}) for the power stage (to optimize efficiency and output power).

Another crucial issue, especially in this technology platform, is related to the parasitics of the connection between the output balun, T_{OUT}, and the ground pads. Such a connection is typically implemented by means of a metal ground plane (GP) and is modeled by a lumped inductance, L_{GP}, and a resistance, R_{GP}, placed at the secondary coil of T_{OUT} (see Figure 7.21).

Actually, the value of L_{GP} is related to the ground path length and is usually in the range of 25 to 50% (depending on the BEOL of the process and metal density rules) of the transformer inductances. Of course, the choice of a suitable structure for GP is highly related to the adopted BEOL. In general, different and conflicting requirements have to be balanced. Multilayer stacked structures achieve low-resistance/low-inductance ground paths, thanks to the increased equivalent thickness. On the other hand, the use of lower metals produces a detrimental increment of the parasitic capacitances toward the substrate. Moreover, due to the metal density rules, proper holed planes have to be used for each metal layer. Possible arrangements for a metal GP exploiting the highest Cu metals (metal 6, metal 5, and metal 4) are shown in Figure 7.22. From the top to the bottom, the structures differ for the number of metal layers and exhibit decreasing specific inductance/resistance values at the cost of higher parasitic capacitances. In the adopted technology, the three-layer structure gives better results and is used to cover all the unused chip area. The EM simulations of the GP path at the PA output returned L_{GP} and R_{GP} at 77 GHz of around 30 pH and 0.9 Ω, respectively. Such values would cause a considerable reduction of both k and Q-factor of T_{OUT} and thus a degradation of the PA performance. This drawback is overcome by shunting the GP connection at the operating frequency through the resonant LC series network (L_R, C_R), which is used to cancel out the impedance toward ground at the secondary coil of balun T_{OUT} (node X).

Compared with trivial inductance cancellation (e.g., by means of series capacitor or output matching networks), the adopted GP cancellation technique has the advantage of also removing the GP resistive contribution (R_{GP}) that is highly detrimental for PA performance. Indeed, the residual parasitic effect of the L_RC_R networks (i.e., R_R in Figure 7.21) can be properly minimized

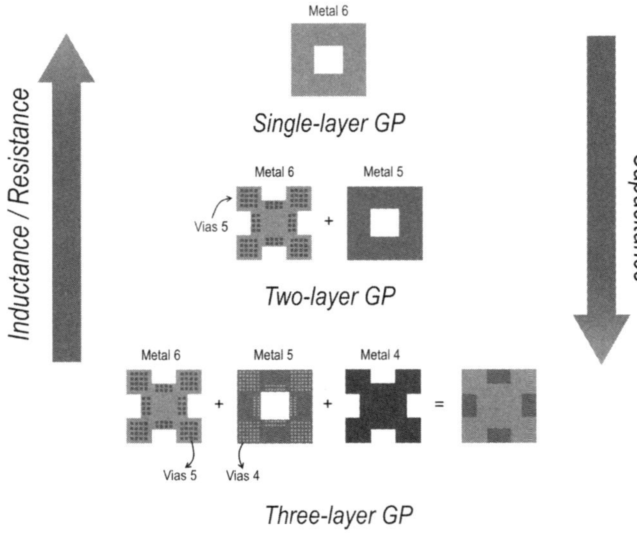

Figure 7.22 Examples of different metal GP structures.

by using a high-Q multilayer microstrip (L_R = 65 pH, R_R = 0.15 Ω at 77 GHz) directly connected to the ground pad. The resonance is easily set by means of a MIM capacitor of 65 fF.

Finally, some words should be said about integrated transformer design (T_{IN}, T_{LO}, T_1, T_2, and T_{OUT}), whose structures are depicted in Figure 7.21. In the proposed LRR TX, transformers are designed to operate at IF, LO, and millimeter-wave frequencies (i.e., 25.5 GHz, 51 GHz, and 76.6 GHz, respectively) and hence different inductance values (i.e., from 50 to 400 pH), configurations (i.e., stacked or interleaved), and coil geometries (i.e., single or multiturn coils, spiral or folded microtrip coils) are used. Table 7.4 summarizes geometrical and electrical features of adopted transformers. The BiCMOS9 metal stack is the same already reported in Figure 7.17.

Transformer T_{IN} exploits an interleaved configuration with 1:2 turn ratio to implement the 50-Ω impedance matching at the IF port. The other transformers take advantage of stacked two-layer configurations to maximize *k*. Coil widths are optimized to trade off series resistance and self-resonance frequency. It is worth noting that millimeter-wave transformers (i.e., T_1, T_2, and T_{OUT}) do not exploit a PGS since the benefits at such high frequency are negligible, while it considerably increases the capacitive parasitic and hence reduces the resonance frequency.

7.3.2 Experimental Results of the 77-GHz TX Front-End

The die micrograph of the 77-GHz LRR TX front-end is shown in Figure 7.23. The die size is 800 μm × 725 μm, including on-chip interface baluns. All measurements are performed at a standard 2.5V supply voltage.

Output power and power gain of the 77-GHz TX as a function of the input power at 25.5 GHz are reported in Figure 7.24(a). The TX power

Table 7.4
Transformer Geometry and Electrical Data

	T_{IN} Freq = 25.5 GHz		T_{LO} Freq = 51 GHz		$T_1 - T_2 - T_{OUT}$ Freq = 76.5 GHz	
Transformer	**Primary (P)**	**Secondary (S)**	**Primary (P)**	**Secondary (S)**	**Primary (P)**	**Secondary (S)**
Metal layers	Al-M6-M5	Al-M6-M5	Al-M6	M5-M4	Al-M6-M5	Al-M6-M5
Structure	Interleaved 1:2		Stacked		Stacked folded microstrip	
PGS	YES		YES		NO	
d_{in} [μm]	48		56		15	
w [μm]	6	6	5	5	5 - 5 - 8	5 - 5 - 8
L [pH]	150	390	140	140	71 - 82 - 55	71 - 82 - 55
Q-factor	21	20	22	19	21 - 20 - 18	21 - 20 - 18
k	0.7		0.75		0.76 - 0.79 - 0.76	

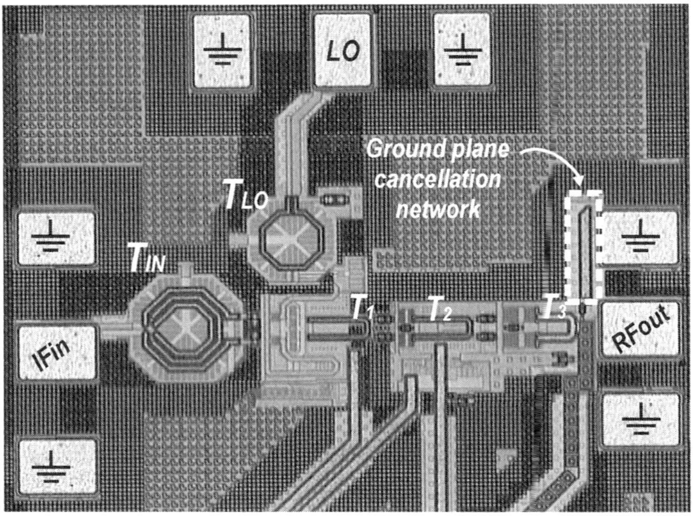

Figure 7.23 Die photo detail of the LRR TX. (© 2010 IEEE. Reprinted with permission from [12].)

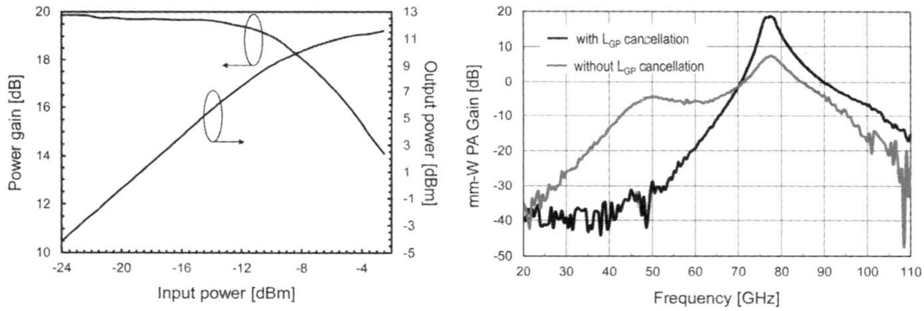

(a) (b)

Figure 7.24 Measured performance: (a) TX output power and power, and (b) comparison of measured gain of millimeter-wave PAs. (Source: [27]. Reprinted with permission.)

conversion gain, the maximum output power, and the output-referred 1-dB compression point (OIP_{1dB}) are 20 dB, 12 dBm, and 9.5 dBm, respectively. The 77-GHz TX draws an overall current of 158 mA from 2.5V supply voltage.

Large-signal measurements have also been performed for the stand-alone 77-GHz PA, which exhibits a maximum gain of 18.5 dB and an output power of 12 dBm with a power-added efficiency (PAE) of 4.7% at 77 GHz. This achievement is mainly due to the adoption of the GP cancellation technique. Indeed, this approach allows a 10-dB improvement in the power gain. Figure 7.24(b) compares the measured gain performance of stand-alone PA with and without the GP cancellation network. It is worth noting that no undesirable effects are produced in a wide frequency range, thus confirming the reliability of the proposed cancellation technique.

Despite the relatively low f_T/f_{max}, the 77-GHz TX demonstrates state-of-the-art performance in terms of both gain and output power. Moreover, the adoption of a single PLL for dual-band operation allows considerable saving of both current consumption and silicon area.

7.4 W-Band TX Front-End for FMCW Automotive Radar

A millimeter-wave optimized process [2] (i.e., BiCMOSMW) is here exploited to implement a W-band TX front-end for FMCW automotive radar complying with typical LR requirements. The most challenging specification for a radar transmitter is the output power level. A power level as high as 14 dBm is required to cover detection ranges higher than 100m. For instance, assuming a radar cross section of 30 m^2 at 77 GHz (i.e., corresponding to a car), typical antenna gain and bumper loss of 18 dBi and 3 dB, respectively, the maximum detection range is about 120m, provided that the saturated output power is at least 14 dBm and the receiver sensitivity is better than –105 dBm. Another key specification of a radar transmitter is the phase noise coming from the LO since it considerably impacts on the IF noise density at the RX output (an increase of 10 dB in phase noise worsens the IF noise density at the RX output by 5 dB) [16]. This produces a degradation of radar sensor capability of detecting a small/far target hidden by a bigger/close one. Typical phase noise values requested by the LR application are around –95 dBc at 1-MHz frequency offset. Finally, to account for process/voltage/temperature (PVT) variations, a wide tuning range for the VCO is highly recommended.

Based on these basic considerations, design efforts have to be mainly focused on the crucial blocks of a W-band transmitter for FMCW radar (i.e., the PA and the VCO).

Figure 7.25 depicts the block diagram of the TX front-end, which adopts a PLL-based scheme to drive the millimeter-wave PA. The W-band PLL exploits a well-established architecture composed of a PFD, a CP, an on-chip loop filter, a 77-GHz VCO, and N-integer divider. An external reference (17.70–18.55 MHz) provided by a digital direct synthesizer (DDS) along with the divider by 4096 allows a frequency sweep from the 76 to 77 GHz to be achieved. The PLL also generates a low frequency signal (f_0/16) for testing purpose (e.g., VCO measurement).

7.4.1 Design of the W-Band TX Front-End

Innovative circuit/layout solutions are adopted for both VCO and PA to fulfill phase noise and output power specifications, respectively. Figure 7.26 shows

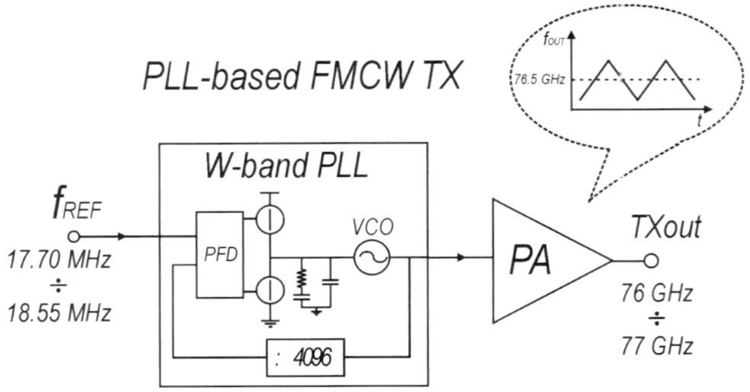

Figure 7.25 Block diagram of the W-band TX front-end for FMCW automotive radar.

the schematic of the W-band TX core, consisting of a differential Colpitts-based VCO (closed in a PLL) that drives a two-stage PA by means of common-collector buffers.

As far as the VCO is concerned, the attainment of low phase noise and wide tuning range, together with a high-power oscillation signal to properly drive the PA, is not trivial at W-band frequencies. Phase noise and tuning range progressively degrade at increasing operating frequencies due to lower quality factor of varactors and higher contribution of parasitic capacitances, respectively. Usually, a design tradeoff between these two parameters is also required.

First-rate performance for a W-band VCO is achieved by taking advantage of an effective solution to provide the oscillation signal to the output buffer [18, 42]. Specifically, the VCO signal, V_{out}, is drawn directly from the switching base terminals of transistors Q_1, Q_2 by means of an integrated transformer, T_B, whose primary winding performs the resonant inductor while the secondary winding is used to deliver the oscillation signal to the output buffers. This topology differs from a traditional Colpitts VCO for millimeter-wave applications [43, 44], which typically includes load inductances in the switching collector terminals to provide the oscillation signal to an output buffer. Such a traditional approach is justified at RF due to the large base-to-collector current gain (i.e., β_{ac}), but it becomes less convenient at W-band frequencies since oscillators operate close to f_T, and β_{ac} can greatly be degraded depending on collector load condition and voltage swing. The elimination of the load inductors produces a significant improvement of the oscillation signal swing. Specifically, the increase in the transistor swing out of the saturation region allows extending the currently limited regime in which the phase noise decreases with the increase of the oscillation amplitude that is proportional to the bias current.

In the adopted Colpitts-based VCO, the inductive component of the resonant network is implemented by the primary coil of transformer T_B, as mentioned before, whereas the resonant capacitor, C_{VAR}, is implemented by

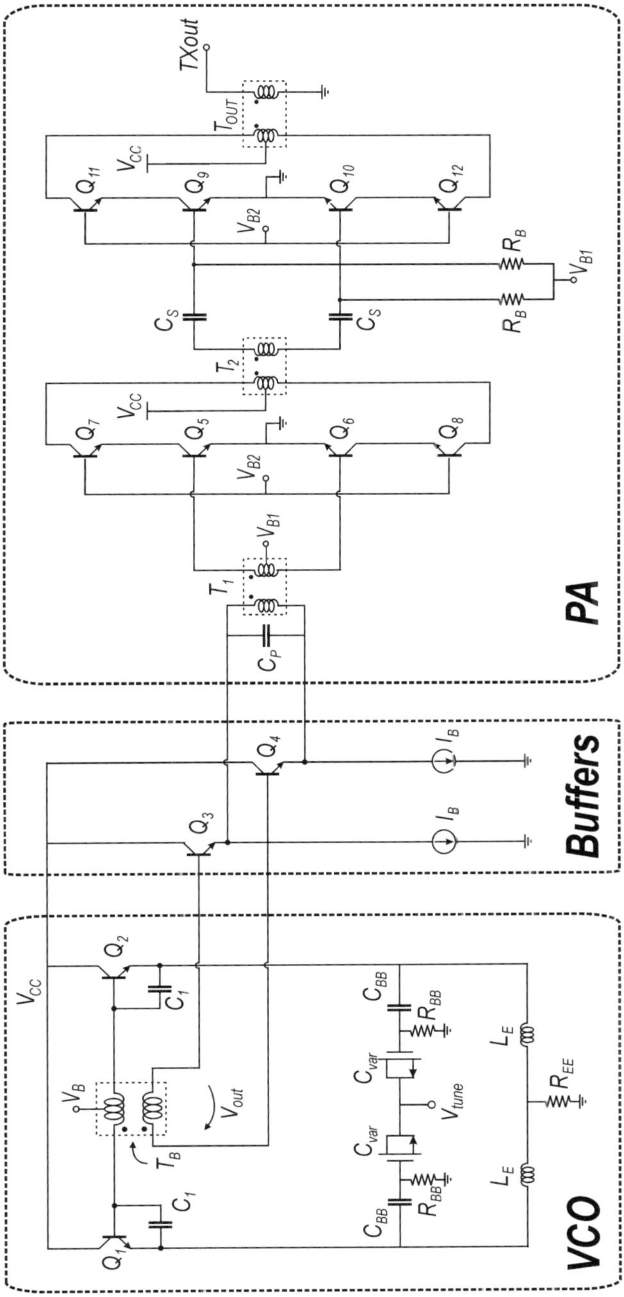

Figure 7.26 Schematic of the W-band TX core.

varactors with a single-ended tuning voltage control (V_{TUNE}) to simplify the design of the PFD/CP. To exploit the whole positive tuning characteristic of the varactors and reduce the effect of parasitic capacitances, they are biased by means of R_{BB} and C_{BB} (i.e., the oscillation signal is ac-coupled). Inductors L_E are included to provide a high-impedance path between the emitters. Common-mode resistor R_{EE} is adopted to prevent noise coming from additional active devices (i.e., tail current mirror), while setting the emitter voltage to avoid collector-emitter breakdown of Q_1-Q_2. Finally, the VCO bias current is accurately set by means of a pseudo-mirror scheme (not included in Figure 7.26) through the center-tap of T_B (i.e., V_B).

It is worth noting that for a traditional Colpitts-based VCO, the oscillation frequency is given by the following expression:

$$f_0 = \frac{1}{2\pi\sqrt{L_B \times \left[(1+A)\times C_{bc} + \frac{C_{BE}C_{VAR}}{C_{BE}+C_{VAR}}\right]}} \quad (7.7)$$

where C_{BE} is the base-emitter capacitance and accounts for C_1, C_{bc} is the base-collector capacitance and is multiplied by the Miller effect factor, (1+A), as –A is the inverting gain from the base to the collector. It is apparent that the tuning action of C_{VAR} on f_0 is reduced by the fixed term due to C_{bc}. For this reason, several circuit arrangements were proposed in the literature, such as cascode topologies [43] (that unfortunately require higher supply voltages) or Miller-effect cancellation networks [43] (that exhibit not negligible sensitivity). On the other hand, the solution in Figure 7.26 has the advantage of inherently eliminating the Miller effect (i.e., $A = 0$) with consequent benefits in terms of tuning range performance. In turn, the increased tuning range can be exchanged with smaller (and hence higher-Q) varactors or with a larger inductance, thus reducing phase noise or current consumption, respectively.

Besides the increased oscillation signal swing and the elimination of the Miller effect on C_{bc}, another benefit of this VCO topology is the high isolation provided by the transformer coupling. Indeed, T_B provides high common-mode isolation from the output buffers, which is not achievable by using either dc or capacitive coupling at the collectors.

Regardless of the arrangements of the VCO with respect to more traditional topologies, the implementation of a low-noise oscillator follows well-known design guidelines [43, 44]. Especially at millimeter-wave frequencies, a crucial point is the maximization of the LC tank Q-factor, which is limited by the varactors. In this design, A-MOS varactors with minimum-length/minimum-width are used to take advantage of the better Q-factor of elementary devices. Unfortunately, the varactor Q-factor at 77 GHz is lower than 10 in the adopted technology, and the tank inductor design is constrained by both Q-factor and oscillation amplitude. According to Leeson's formula, the minimization of the phase noise requires a resonant tank with high capacitance and low inductance. On the other hand, high oscillation amplitude and low current consumption need high inductor values. Finally, the inductor value imposes the total varactor size and hence the tuning range performance. Therefore, all these aspects have to be taken into account to set the resonator components.

As far as the switching pair Q_1-Q_2 is concerned, transistors are biased at a current density just below the peak-f_T. Moreover, base resistance is minimized by adopting multiple shunt devices. The reduction of both external base resistance and parasitic capacitances is obtained by using as elementary device a simple double-base BJT instead of a multifinger transistor.

According to simulations, the estimated performance of phase noise and tuning range are –99 dBc/Hz at 1-MHz offset and 5 GHz, respectively. The estimated output power of the VCO is around 2 dBm, which is high enough to drive the PA into saturation.

Let's now compare Figures 7.26 and 7.21.The comparison highlights that the 77-GHz PA adopts the same topology already exploited for the previous design in BiCMOS9 technology (i.e., a two-stage transformer-coupled pseudo-differential topology). Of course, the amplifier in Figure 7.26 takes advantage of the higher f_T/f_{max} and the millimeter-wave optimized BEOL of the BiCMOSMW process (see Figure 7.1). Due to technology advancements, some design changes should be considered, as listed next:

1. Thanks to the improved f_T/f_{MAX}, the traditional design procedure for millimeter-wave two-stage PAs [15, 41] is adopted. Indeed, the bias current density is set at the peak f_T/f_{MAX} for the driver stage to maximize the small-signal power gain, whereas a reduced current density (i.e., 50% of peak f_T/f_{MAX}) is used for the power stage to optimize the PAE and the output power.
2. Thanks to the improved BEOL, there is no need to include any GP cancellation network at the output of the PA.
3. The increased thickness of both metal 6 and intermetal oxides allows a single-layer GP to be used (see Figure 7.22).
4. Spiral transformers are used in place of stacked folded microstrips adopted in the previous design.
5. The layout floor plan/routing for the elementary CAS stage is aimed at the reduction of emitter parasitic inductances and hence gain losses [15].

Thanks to the advanced technology platform and the consequent design changes, the PA is able to considerably improve the performance previously obtained in BiCMOS9 in terms of saturated output power, gain, and efficiency, achieving the simulated values of 15.8 dBm, 23 dB, and 8.4%, respectively.

Some words should be said for the design of the prescaler, which, after the VCO, is the most crucial block in the PLL. It is implemented by four divide-by-two stages. Static dividers are used to take advantage of their high robustness and larger operating frequency range at millimeter-wave frequencies. Each divider adopts ECL flip-flops (i.e., D-latches) in a master–slave configuration, as already adopted in the 24-GHz SRR sensor (see Figure 7.6). Instead of traditional emitter followers, ac coupling is used to interconnect the dividers. Indeed, this solution saves current consumption and allows an optimum bias level for each divider cell to be achieved, without considerably affecting silicon area. Both load and current in the latch are set to provide an output voltage swing from 250 to 350 mV from the first to the last stage. The first divider is loaded with an R-L series impedance to extend bandwidth at a reduced current budget thanks to the inductive peaking technique, whereas following dividers adopt the traditional resistive load. The signal $f_0/16$ produced at the prescaler output is used to easily perform the VCO test at low frequency.

As already highlighted, an important role in the millimeter-wave design is played by integrated transformers (i.e., T_B, T_1, T_2, and T_{OUT} in Figure 7.26). Three-dimensional views of adopted transformers, along with the corresponding metal stack, are shown in Figure 7.27, while their geometrical/electrical features are summarized in Table 7.5. Transformer T_B is implemented by means

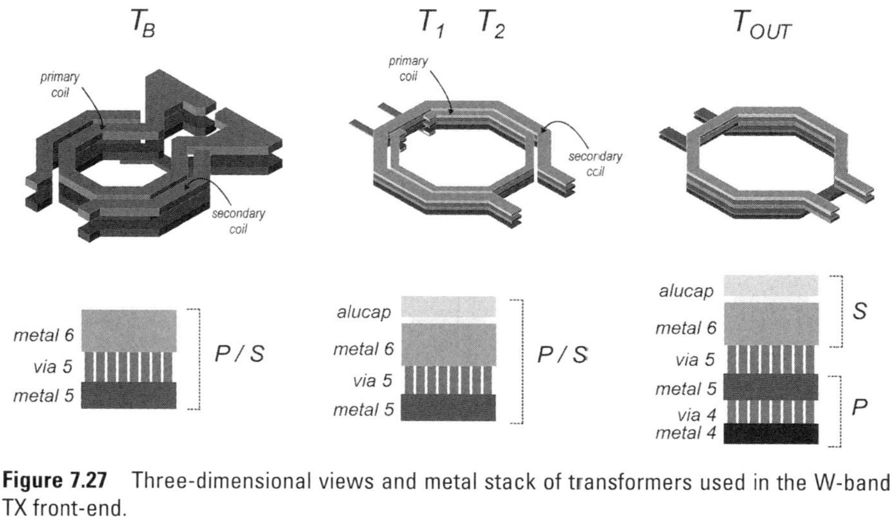

Figure 7.27 Three-dimensional views and metal stack of transformers used in the W-band TX front-end.

Table 7.5
Transformer Geometry and Electrical Data at 77 GHz

	T_B		$T_1 - T_2$		T_{OUT}	
Transformer	**Primary (P)**	**Secondary (S)**	**Primary (P)**	**Secondary (S)**	**Primary (P)**	**Secondary (S)**
Metal layers	M6-M5	M6-M5	Al-M6-M5	Al-M6-M5	Al-M6	M5-M4
Structure	Interleaved		Interleaved		Stacked	
PGS	NO		NO		NO	
d_{in} [μm]	30		25–30		30	
w [μm]	5	5	5	5	5	5
L [pH]	56	67	50–65	75–80	65	65
Q-factor	20	19	17.5–18	16.5–20	16	16
k	0.45		0.42		0.55	

of an interleaved topology. Both primary and secondary coils exploit a two-layer structure (i.e., metal 6 and metal 5), while lower metal 4 is used only as underpass. A spiral width of 5 μm is used to reduce the parasitic capacitances toward the substrate without affecting winding Q-factor. The coil spacing is set to the minimum value to improve the magnetic coupling that is involved in the oscillation signal transfer. The layout of T_B inherently helps the interconnections of the VCO core, thus minimizing RLC parasitics.

An interleaved structure is also adopted for the PA input and intermatching transformers (T_1 and T_2). For both T_1 and T_2, a three-layer structure allows spiral series losses to be decreased at the cost of a slight reduction of inductance (due to the increased equivalent metal thickness). A stacked configuration is instead preferred for T_{OUT} to obtain a slightly higher k (i.e., 0.55 against 0.42). In this case, transformer windings consist of a two-layer structure (i.e., alucap/metal6 and metal5/metal4). By comparing Tables 7.4 and 7.5, stacked structures in BiCMOSMW exhibit lower magnetic coupling factors. This is a drawback due to the increased thickness of the oxide layer between metal 6 and metal 5 (see Figure 7.1) in the BiCMOSMW technology. It can be overcome by using advanced transformer structures that exploit mixed interleaved/stacked magnetic coupling factors [45]. Narrow coils of 5 μm are also used for T_1, T_2, and T_{OUT} in order to reduce the parasitic capacitances, thus allowing self-resonance frequency above 150 GHz. Finally, polysilicon or metal PGS structures are not adopted at such high frequencies, as already said.

For all the transformers, the distance between spirals and GP is set to 30 μm to avoid the rising of loss effects due to magnetic couplings with the metal plane. This choice also allows density rules for chemical/mechanical planarization (CMP) integrity to be fulfilled since a 30-μm dummy-free zone around the spiral can be tolerated in the final IC.

7.4.2 Experimental Results of the W-Band TX Front-End

The die micrograph of the TX front-end is shown in Figure 7.28. The PLL frequency reference and the TX output are placed on the left- and the right-hand side, respectively. The die size is 1240 μm × 1310 μm. Macroblocks are highlighted for the sake of clarity. All measurements are performed at a standard 2.5V supply voltage.

The VCO oscillation frequency, f_o, is shown as a function of the tuning voltage, V_{tune}, in Figure 7.29. With V_{tune} sweeping from 0 to 2.5V, the tuning range is of about 4.9 GHz (from 74.4 to 79.3 GHz), which is large enough to guarantee the 76–77 GHz operation in presence of PVT tolerances. The phase noise measurement at 4.75 GHz (i.e., at $f_o/16$) is displayed in Figure 7.30. The VCO is compliant with the phase noise specification since it achieves a value of –99.3 dBc/Hz at 1-MHz offset from the 76-GHz carrier. The spectrum of the low-frequency divider output at $f_o/16$ is also shown in Figure 7.31 to highlight the lack of significant spurs.

The millimeter-wave PA is also evaluated as a stand-alone block [15]. The measured large signal parameters PA are reported in Figure 7.32. The PA saturated output power is 15 dBm with a peak PAE of 7.5%. The quiescent current is 130 mA. The OIP_{1dB} is 12 dBm. The small-signal power gain at 77 GHz is higher than 22 dB.

The TX power capability is demonstrated in Figure 7.33, which depicts the TX output spectrum. The measured power of –52 dBm has to be de-embedded for the test bench loss (i.e., about 67 dB). Therefore, the transmitter front-end is able to deliver an output power of 15 dBm, confirming that the PA is well driven into saturation by the VCO. The achieved output power

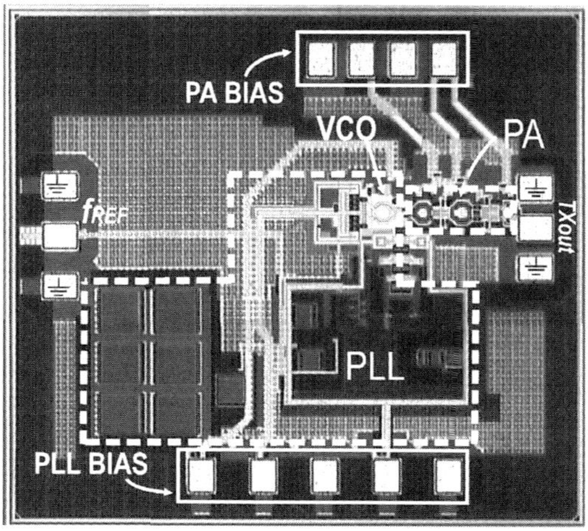

Figure 7.28 Die photograph of the W-band TX front-end.

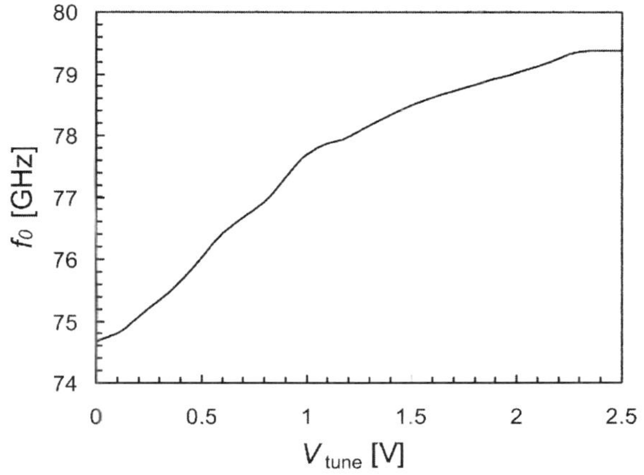

Figure 7.29 VCO oscillation frequency (f_0) versus tuning voltage V_{tune}.

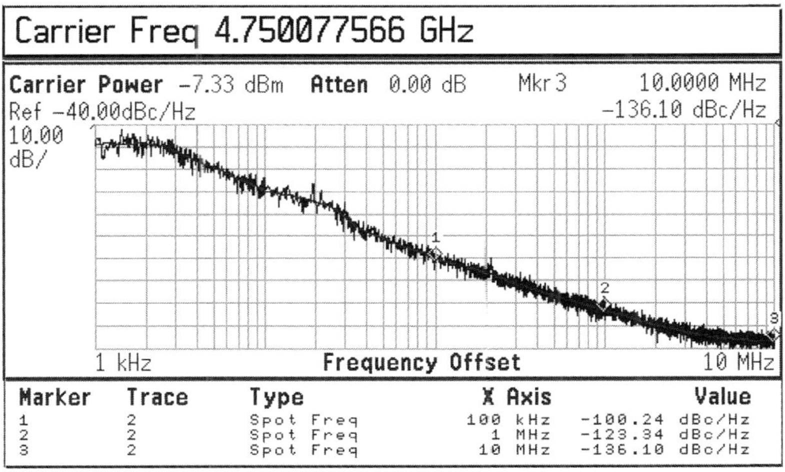

Figure 7.30 VCO phase noise performance measured at $f_0/16$ = 4.75 GHz. (© 2013 IEEE. Reprinted with permission from [18].)

performance is among the best results achieved by 77-GHz SiGe bipolar integrated circuits at that time [16, 46].

7.5 W-Band RX Front-End for FMCW Automotive Radar

The last section of the chapter is dedicated to the implementation of a W-band RX front-end for FMCW automotive radar in BiCMOSMW technology [2]. It can be useful spending some time to highlight how the performance of a

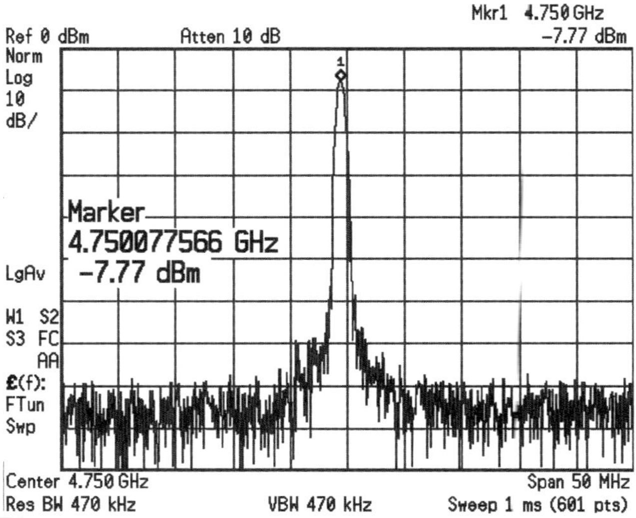

Figure 7.31 Frequency spectrum measured at divider output (f_0 = 76 GHz).

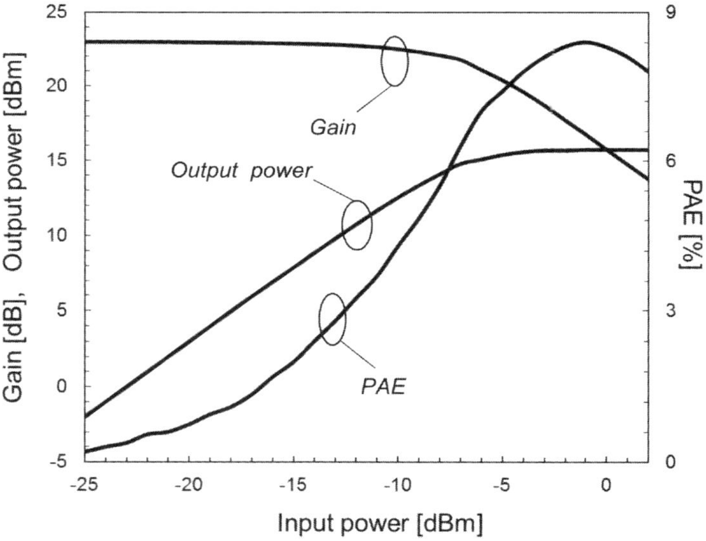

Figure 7.32 Measured large-signal parameters of the PA.

FMCW radar system is related to the RX characteristics. Assuming conjugate matching at the antenna, the SNR of a radar system in a first approximation is given by the well-known expression

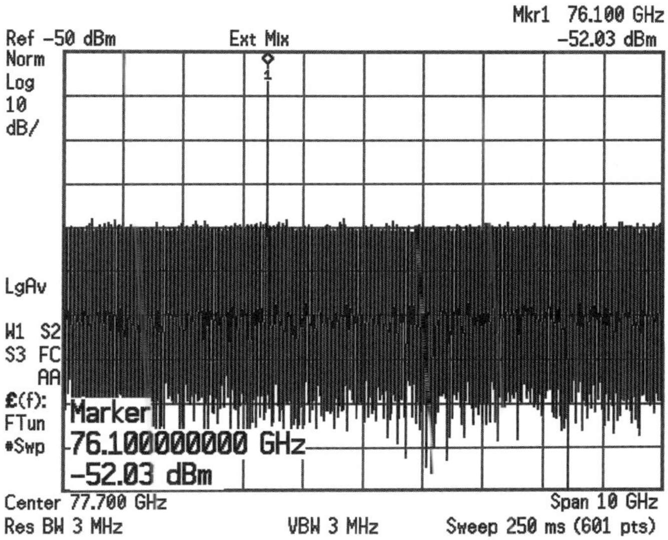

Figure 7.33 Measured output spectrum of the W-band TX. (© 2013 IEEE. Reprinted with permission from [18].)

$$
\begin{aligned}
SNR\ [\mathrm{dB}] &= P_{OUT}^{TX}[\mathrm{dBm}] - L\ [\mathrm{dB}] \\
&+174\ [\mathrm{dBm/Hz}] - NF_{RX}[\mathrm{dB}] - 10 \cdot \log(B)
\end{aligned}
\tag{7.8}
$$

where $P_{\mathrm{OUT}}^{\mathrm{TX}}$ is the output power of the transmitter, L is the loss term, –174 dBm/Hz is the thermal noise density (i.e., 10log(kT/1mW)), NF_{RX} is the noise figure of the receiver, and B is the bandwidth of the system. For the sake of completeness, term L summarily accounts various loss contributions besides the path loss (i.e., some parameters of the radar equation such as antenna gain, target radar cross-section, and so on), further terms due to the relative target position with respect to the beam axis, and finally contributions due to the baseband signal processing. It is worth noting that (7.8) does not depend on the conversion gain of the receiver, which is related to the noise floor at the IF output.

As clearly shown by (7.8), the RX key parameter is the noise figure, NR_{RX}, whose minimization is mandatory to reduce the TX output power and consequently the power consumption. This is even more important for radar systems, which are affected by high-power reflections of the FMCW signal due to the radome and/or the car bumper. Therefore, high output power turns out in very challenging linearity specifications for the RX chain, which can be mitigated provided that a better noise performance is guaranteed.

Based on these considerations, the key RX block of a FMCW radar front-end is the down-conversion mixer that should exhibit very demanding IIP_{1dB} while maintaining low noise figure. For these reasons, significant efforts have been addressed in the last years toward the development of highly linear mixer topologies [47–50]. Moreover, several silicon radar implementations [11, 16, 49] waive the LNA stage before the mixer to maximize the RX linearity performance, thus achieving lower gain at millimeter-wave frequency and hence higher noise.

Figure 7.34 depicts a simplified block diagram of the proposed RX radar front-end. As with the TX front-end described in Section 7.4, it adopts a PLL-based scheme driven by an external DDS. Since the W-band PLL was already described, here the attention will be mainly focused on the RX chain. It includes an LNA and a down-conversion mixer (i.e., LNA/mixer). Given a P_{OUT}^{TX} of 15 dBm (as the one delivered by the W-band TX front-end described in Section 7.4), the LNA/mixer should exhibit a noise figure lower than 12 dB. Moreover, an IIP_{1dB} around –10 dBm should be provided to preserve sensitivity in presence of high-power reflections. Finally, to avoid noise figure degradation due to baseband circuitry, a conversion gain around 15–20 dB has to be guaranteed.

7.5.1 Design of the W-Band RX Front-End

The fulfillment of stringent linearity specifications along with low noise figure and adequate conversion gain requires the adoption of proper circuit topologies for the mixer stage. Traditionally, the problem has been faced using Gilbert-

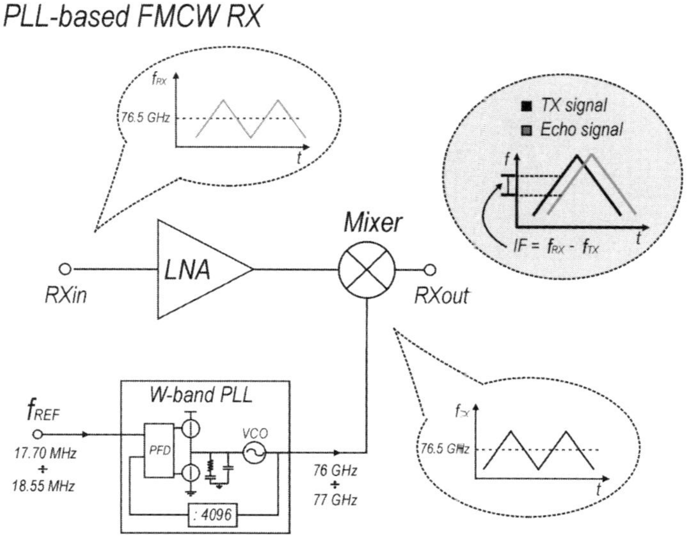

Figure 7.34 Block diagram of the W-band RX front-end for FMCW automotive radar.

quad based mixer topologies operated at high supply voltages (i.e., 3.3V or even 5V) and high currents to reduce the output voltage clipping and improve the linearity of the V/I converter, respectively [47, 48]. On the other hand, the use of high currents is detrimental for the noise contribution of the Gilbert quad (due to the increasing shot noise translated to the IF by the switching quad operation), thus affecting the overall noise figure. For these reasons, current-steering solutions are usually adopted to properly bias the V/I converter and the quad.

An alternative yet effective approach to trade off mixer linearity and noise is to adopt inherently linear configurations based on a current-driven Gilbert quad. The proposed solution exploits a transformer-driven Gilbert quad, in a similar fashion of the one exploited for the 24-GHz mixer described in Section 7.2 (see Figure 7.8). A similar solution has been recently adopted for a W-band radar receiver [16] demonstrating at 3.3V supply voltage an IIP_{1dB} of –5 dBm with a noise figure and a conversion gain of 10 dB and 19 dB, respectively.

In our implementation, the key block is a highly linear mixer cell (i.e., IIP_{1dB} better than 0 dBm) operating at a supply voltage as low as 2.5V. An input amplifier (LNA) is also profitably included to improve the overall RX conversion gain at the expense of an acceptable degradation of the IIP_{1dB}, but guaranteeing better input/noise matching.

The simplified schematic of the 77-GHz LNA/mixer is shown in Figure 7.35. At the input, an integrated transformer, T_{in}, performs the single-ended-to differential conversion of the RF signal, while delivering it to the emitters of a differential CB amplifier, Q_1-Q_2. T_{in} is also used to achieve 50 Ω input matching operating in resonant mode with capacitors C_{S1} and C_{P1} to reduce its IL. The current density of the CB pair is properly chosen to minimize noise, while the emitter area determines the noise optimum source impedance. Resistor R_{EE} is sized to guarantee that transistors work within the BV_{CE} limit. The amplifier load is implemented by an integrated transformer, T_{LOAD}, whose TCR—see (7.6)—has been sized to obtain a LNA gain of around 10 dB. T_{LOAD} also performs the V-I conversion of the output signal toward the Gilbert quad. The adopted transformer coupling between the amplifier and the mixer also allows noise reduction since the bias current used for the quad is lower than that used in the input stage (i.e., 6 mA and 14 mA, respectively). The Gilbert quad is driven by the LO signal coming from the VCO buffers by means of on-chip transformer T_{LO}, whose central tap is exploited for biasing according to a mirror-like scheme. Thanks to the biasing solution and the LO driving provided by T_{LO}, the quad works at the maximum voltage swing (i.e., around 1.6V) and its noise performance at high input signal levels is improved. Finally, a resistive load is used, which guarantees a wideband matching on a 100 Ω differential load.

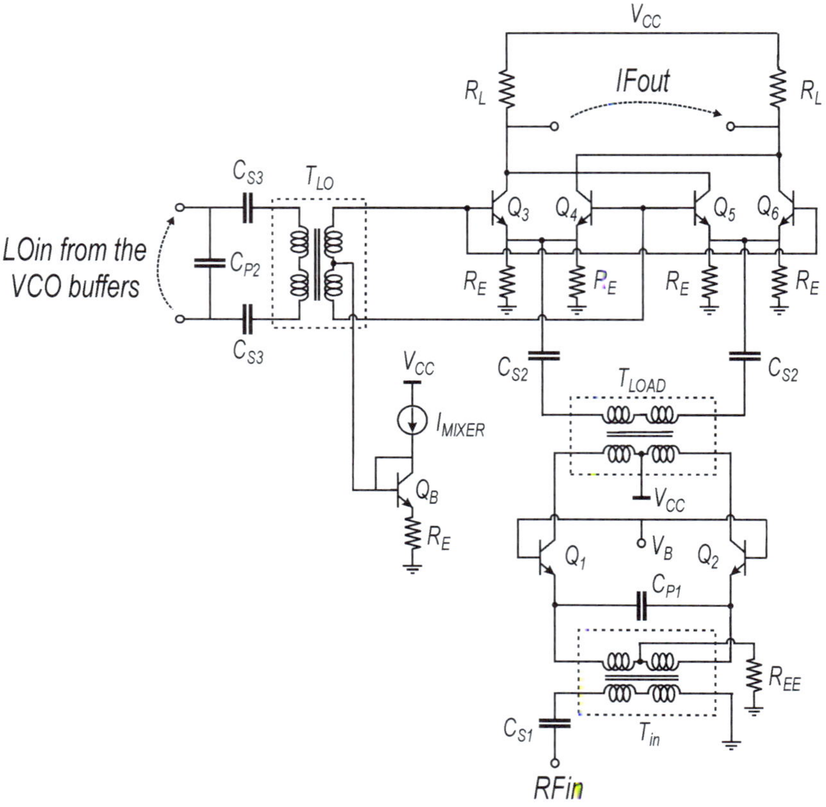

Figure 7.35 Schematic of the W-band LNA/mixer.

All on-chip transformers are implemented by means of an interleaved topology. Both primary and secondary coils exploit a three-layer structure (i.e., alucap , metal 6, and metal 5), while lower metal 4 is used only as underpass. Transformer metal stack is the same as what was already shown in Figure 7.27. For the sake of completeness, transformer geometrical/electrical features are summarized in Table 7.6.

According to simulations, the LNA/mixer achieves a IIP_{1dB} of about –10 dBm with a conversion gain of 17.5 dB and a noise figure of 10.5 dB at 1-MHz IF, while drawing about 20 mA from a 2.5V supply voltage.

7.5.2 Experimental Results of the W-Band RX Front-End

The die micrograph of the RX front-end is shown in Figure 7.36. The 77-GHz input, the PLL frequency reference and the differential IF output are placed on the left-hand side, on the right-hand side, and on the top, respectively. The die size is 1175 μm × 1460 μm. Macroblocks are highlighted for the sake of clarity.

Table 7.6
Transformer Geometry and Electrical Data at 77 GHz

Transformer	T_{IN} Primary (P)	T_{IN} Secondary (S)	T_{LOAD} Primary (P)	T_{LOAD} Secondary (S)	T_{LO} Primary (P)	T_{LO} Secondary (S)
Metal layers	Al-M6-M5	Al-M6-M5	Al-M6-M5	Al-M6-M5	Al-M6-M5	Al-M6-M5
Structure	Interleaved		Interleaved		Interleaved	
PGS	NO		NO		NO	
d_{in} [μm]	44		44		35	
w [μm]	5	5	5	5	5	5
L [pH]	117	113	124	104	98	116
Q-factor	21.4	18.4	19	18.7	12	18.7
k	0.46		0.52		0.47	

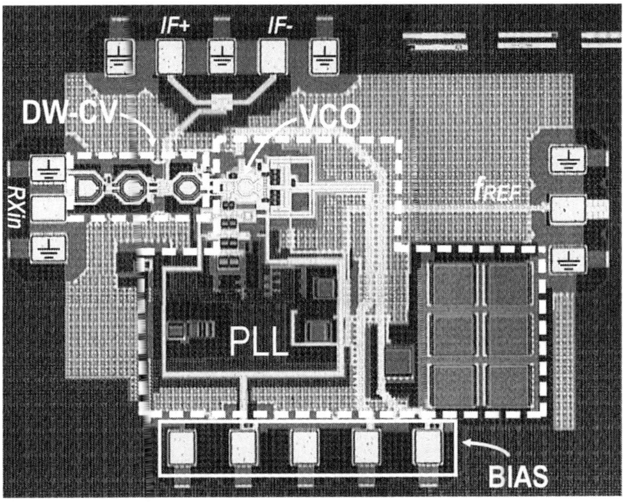

Figure 7.36 Die photograph of the W-band RX front-end.

All measurements are performed at a standard 2.5V supply voltage. The LO power, P_{LO}, provided by the VCO buffers is around 2 dBm, which is suitable to properly drive the quad while avoiding degradation of both the noise figure and conversion gain.

Figure 7.37 depicts the conversion gain and the input matching versus frequency. In the operative frequency range, the LNA/mixer achieves an almost flat conversion gain of 17 dB while maintaining an S_{11} better than –14 dB. As shown in Figure 7.38, the LNA/mixer exhibits a noise figure of 11 dB at 1-MHz IF. The figure also highlights the linearity performance of the RX front-end, which achieves an IIP $_{dB}$ of about –10 dBm. At the compression point, the

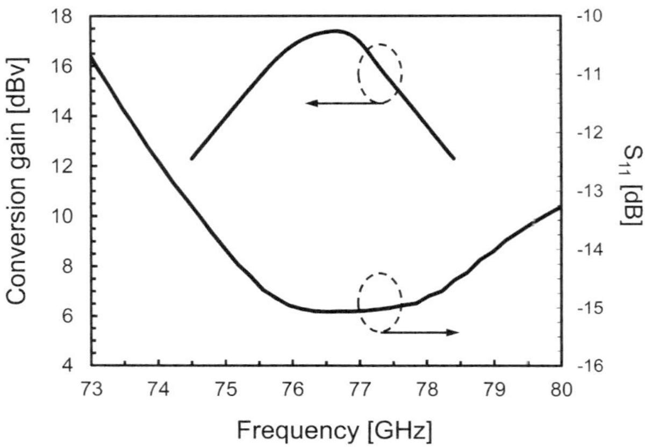

Figure 7.37 Conversion gain and input matching versus frequency (IF = 1 MHz, P_{LO} = 2 dBm).

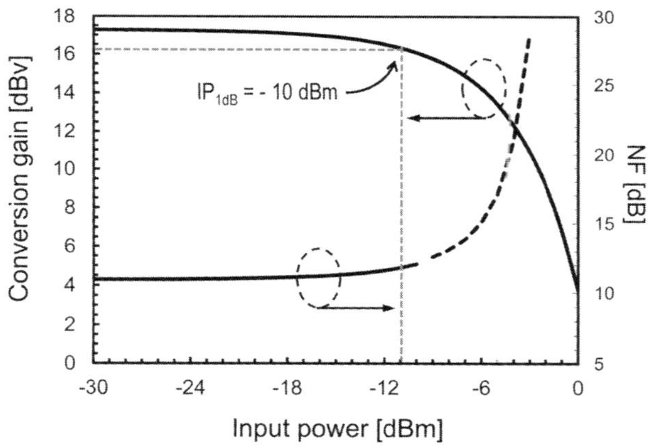

Figure 7.38 Conversion gain and noise figure versus input power (RF = 77 GHz, IF = 1 MHz, P_{LO} = 2 dBm).

noise figure still remains lower than 12 dB, and it increases up to 15 dB for an input power of –6 dBm. Such behavior confirms the robustness of the adopted topology against saturation effects due to high-power reflections (e.g., due to bumper or very close targets).

It is worth noting that by removing the CB input stage, better results in terms of linearity could be achieved at the expense of a reduction of the conversion gain (of around 10 dB) and a consequent degradation of the noise figure performance.

References

[1] Laurens, M., et al. "A 150GHz f_T/f_{max} 0.13 μm SiGe:C BiCMOS Technology," *Proceedings of the IEEE Bipolar/BiCMOS Circuits Technology Meeting*, Oct. 2003, pp. 199–202.

[2] Avenier, G., et al., "0.13 μm SiGe BiCMOS Technology Fully Dedicated to mm-Wave Applications," *IEEE J. Solid-State Circuits*, Vol. 44, No. 9, Sep. 2009, pp. 2312–2321.

[3] Rieh, J.-S., et al. "SiGe HBTs with Cut-Off Frequency of 350 GHz," *Proc. Int. Electron Devices Meeting*, Dec. 2002, pp. 771–774.

[4] Huang, W., et al., "SiGe 77 GHz automotive radar technology," *Proc. IEEE Int. Symp. Circuits Syst.*, May 2007, pp. 1967–1970.

[5] Bock, J., et al. "SiGe Bipolar Technology for Automotive Radar Applications," *Proc. Bipolar/BiCMOS Circuits Technol. Meeting*, Sep. 2004, pp. 84–87.

[6] Scuderi, A., et al., *Integrated Inductors and Transformers: Characterization, Design and Modeling for RF and millimeter-wave Applications*, Boca Raton, FL: CRC/Taylor & Francis, 2010.

[7] Ragonese, E., et al., "A Fully Integrated 24GHz UWB Radar Sensor for Automotive Applications," *IEEE International Solid-State Circuits Conference (ISSCC)*, San Francisco, CA, Feb. 2009, pp. 306–307.

[8] Scuderi, A., E. Ragonese, and G. Palmisano, "24-GHz UWB Transmitter for Vehicular Short-Range Radar Applications," *IET Circuits, Devices & Systems*, Vol. 3, Dec. 2009, pp. 313–321.

[9] Ragonese, E., et al. "A SiGe BiCMOS 24-GHz Receiver Front-End for Automotive Short-Range Radar," *Springer Analog Integrated Circuits and Signal Processing*, Vol. 67, May 2011, pp. 121–130.

[10] Bredendiek, C., et al., "A 24GHz Wideband Single-Channel SiGe Bipolar Transceiver Chip for Monostatic FMCW Radar Systems," *Microwave Integrated Circuits Conference (EuMIC)*, Oct. 2012, pp. 309–312.

[11] Forstner, H.-P., et al., "A 77GHz 4-Channel Automotive Radar Transceiver in SiGe," *IEEE Radio Frequency Integrated Circuits Symp. Dig. (RFIC)*, June 2008, pp. 233–236.

[12] Giammello, V., E. Ragonese, and G. Palmisano, "Transmitter Chipset for 24/77-GHz Automotive Radar Sensors," *IEEE Radio Frequency Integrated Circuits Symp. Dig. (RFIC)*, May 2010, pp. 75–78.

[13] Maurer, L., G. Haider, and G. H. Knapp, "77 GHz SiGe Based Bipolar Transceivers for Automotive Radar Applications: An Industrial Perspective," *IEEE Int. New Circuits and Systems Conference (NEWCAS)*, June 2011, pp. 257–260.

[14] Giammello, V., E. Ragonese, and G. Palmisano, "A 77-GHz PA with Ground-Plane Parasitic Cancellation in a SiGe HBT BiCMOS Technology," *Wiley Microwave and Optical Technology Letters*, Vol. 53, June 2011, pp. 1413–1416.

[15] Giammello, V., E. Ragonese, and G. Palmisano, "A 15-dBm SiGe BiCMOS PA for 77-GHz Automotive Radar," *IEEE Trans. Microwave Theory and Tech.*, Vol. 59, No. 11, Nov. 2011, pp. 2910–2918.

[16] Trotta, S., et al., "An RCP Packaged Transceiver Chipset for Automotive LRR and SRR Systems in SiGe BiCMOS Technology," *IEEE Trans. Microwave Theory and Tech.*, Vol. 60, No.3, Mar. 2012, pp. 778–794.

[17] Giammello, V., E. Ragonese, and G. Palmisano, "A Transformer-Coupling Current-Reuse SiGe HBT Power Amplifier for 77-GHz Automotive Radar," *IEEE Trans. Microwave Theory and Tech.*, Vol. 60, n. 6, June 2012, pp. 1676–1683.

[18] Sapone, G., et al., "A 0.13-μm SiGe BiCMOS Colpitts-Based VCO for W-Band Radar Transmitters," *IEEE Trans. Microwave Theory and Tech.*, Vol. 61, No. 1, Jan. 2013. pp. 185–194.

[19] Issakov, V., et al., "A Compact Low-Power 24 GHz Transceiver for Radar Applications in 0.13μm CMOS," *IEEE International Conference on Microwaves, Communications, Antennas and Electronics Systems, (COMCAS)*, Nov. 2009, pp. 1–5.

[20] Jaemo, Y., et al., "A 24-GHz CMOS UWB Radar Transmitter with Compressed Pulses," *IEEE Trans. Microwave Theory and Tech.*, Vol. 60, No. 4, April 2012, pp. 1117–1125.

[21] Wei-Hsiu, H., et al., "An 18dBm Transmitter Frontend with 29% PAE for 24GHz FMCW Radar Applications," *IEEE Radio Frequency Integrated Circuits Symp.*, June 2012, pp. 137–140.

[22] Kawano, Y., et al., "A 77 GHz Transceiver in 90 nm CMOS," *Proc. IEEE Solid-State Circuits Conf.*, Feb. 2009, pp. 310–311.

[23] Mitomo, T., et al., "A 77 GHz 90 nm CMOS Transceiver for FMCW Radar Applications," *IEEE J. Solid-State Circuits*, Vol. 45, No. 4, Apr. 2010, pp. 928–937.

[24] Y.-A. Li,M.-H. Hung, S.-J. Huang, and J. Lee, "A Fully Integrated 77 GHz FMCW Radar System in 65 nm CMOS," *Proc. IEEE Solid-State Circuits Conf.*, Feb. 2010, pp. 216–217.

[25] Cheonsoo, K., et al., "A CMOS Centric 77GHz Automotive Radar Architecture," *IEEE Radio Frequency Integrated Circuits Symp. (RFIC)*, June 2012, pp. 131–134.

[26] Margomenos, A., "A Comparison of Si CMOS and SiGe BiCMOS Technologies for Automotive Radars," *Proc. IEEE Topical Meeting on Silicon Monolithic Integrated Circuits in RF Systems*, (SiRF), Jan. 2009, pp. 1–4.

[27] Ragonese, E., V. Giammello, and G. Palmisano, "A 24/77-GHz SiGe BiCMOS Transmitter Chipset for Automotive Radar," *Wiley Microwave and Optical Technology Letters*, Vol. 55, No. 4, April 2013, pp. 782–786.

[28] Gresham, I., et al., "Ultra-Wideband Radar Sensors for Short-Range Vehicular Applications," *IEEE Trans. Microw. Theory Tech.*, Vol. 52, No. 9, Sep. 2004, pp. 2105–2122.

[29] ETSI EN 302 288-1, "Electromagnetic Compatibility and Radio Spectrum Matters (ERM); Short Range Devices; Road Transport and Traffic Telematics (RTTT); Short-Range Radar Equipment Operating in the 24 GHz Range; Part 1: Technical Requirements and Methods of Measurement."

[30] Federal Communication Commission, First Report and Order, Revision of Part 15 of the Commission's Rules Regarding Ultra Wideband Transmission Systems, ET Docket 98 153, FCC, Washington, DC, 2002.

[31] Ragonese, A., E., and G. Palmisano, "0.13-mm SiGe BiCMOS Radio Front-End Circuits for 24-GHz Automotive Short-Range Radar Sensors," *Proc. IEEE European Solid-State Circuit Conference (ESSCIRC)*, Sep. 2008, pp. 494–497.

[32] Copani, T., et al. "12-GHz Silicon Bipolar Dual-Conversion Receiver for Digital Satellite Applications," *IEEE J. Solid-State Circuits*, Vol. 40, No. 6, June 2005, pp. 1278–1287.

[33] Girlando, G., E. Ragonese, and G. Palmisano, "Silicon Bipolar LNAs in the X and Ku Bands," *Analog Integrated Circuits and Signal Processing*, Vol. 41, Dec. 2004, pp. 119–127.

[34] Kaukovuori, J., et al., "Analysis and Design of Passive Polyphase Filters," *IEEE Trans. Circuits Syst. I*, Vol. 55, Nov. 2008, pp. 3023–3037.

[35] Giammello, V., et al., "An Offset Compensated Baseband Circuit for Automotive Radar Ultra-Wideband Applications," *Proc. IEEE Int. Conf. on Electronics, Circuits and Systems*, Dec. 2007, pp. 931–934.

[36] Yue, C. P., and S. S. Wong, "On-Chip Spiral Inductors with Patterned Ground Shields for Si-Based RF ICs," *IEEE Journal of Solid-State Circuits*, Vol. 33, No.5, May 1998, pp. 743–752.

[37] Dickson, T. O., M.-A. LaCroix, S. Boret, and D. Gloria, et al., "30-100-GHz Inductors and Transformers for Millimeter-Wave (Bi)CMOS Integrated Circuits," *IEEE Transactions on Microwave Theory and Techniques*, Vol. 53, No.1, Jan. 2005, pp. 123–133.

[38] Biondi, T., et al., "Sub-nH Inductor Modeling for RF IC Design," *IEEE Microwave and Wireless Components Letters*, Vol. 15, 2005, pp. 922–924.

[39] Italia, A., et al. "The Transformer Characteristic Resistance and Its Application to the Performance Analysis of Silicon Integrated Transformers," in *IEEE Radio Frequency Integrated Circuits Symp. Dig. (RFIC)*, Jun. 2005, pp. 597–600.

[40] Ragonese, E., et al. "Scalable Lumped Modeling of Single-Ended and Differential Inductors for RF IC Design," *Wiley International Journal of RF and Microwave Computer-Aided Engineering*, Vol. 19, Jan. 2009, pp. 110–119.

[41] Nicolson, S. T., et al. "A Low-Voltage SiGe BiCMOS 77-GHz Automotive Radar Chipset, *IEEE Transactions on Microwave Theory and Techniques*, Vol. 56, No.5, May 2008, pp. 1092–1103.

[42] Sapone, G., A. Italia, E. Ragonese, and G. Palmisano, "Millimeter Wave Oscillator," US patent US8489054 (B2), granted 16 July 2013.

[43] Li, H., and H.-M. Rein, "Millimeter-Wave VCOs with Wide Tuning Range and Low Phase Noise, Fully Integrated in a SiGe Bipolar Production Technology," *IEEE J. Solid-State Circuits*, Vol. 38, No. 2, Feb. 2003, pp. 184–191.

[44] Nicolson, S. T., et al. "Design and Scaling of W-Band SiGe BiCMOS VCOs," *IEEE J. Solid-State Circuits*, Vol. 42, No. 9, Sep. 2007, pp. 1821–1833.

[45] Ragonese, E., G. Sapone, V. Giammello, and G. Palmisano, "Analysis and Modeling of Interstacked Transformers for mm-Wave Applications," *Springer Analog Integrated Circuits and Signal Processing*, Vol. 72, No. 6, June 2012, pp. 121–128.

[46] Li, H., H.-M. Rein, T. Suttorp, and J. Böck, "Fully Integrated SiGe VCOs with Powerful Output Buffer for 77-GHz Automotive Radar Systems and Applications Around 100 GHz," *IEEE J. Solid-State Circuits*, Vol. 39, No. 10, Oct. 2004, pp. 1650–1658.

[47] Dehlink, B., et al., "A Highly Linear SiGe Double-Balanced Mixer for 77 GHz Automotive Radar Applications," in *IEEE Radio Frequency Integrated Circuits Symp. Dig. (RFIC)*, June 2006.

[48] Trotta, S., et al., "Design Considerations for Low-Noise, Highly Linear Millimeter-Wave Mixers in SiGe Bipolar Technology," in *Proc. IEEE European Solid-State Circuit Conference (ESSCIRC)*, Sep. 2007, pp. 356–359.

[49] Trotta, S., et al., "A Multi-Channel Rx for 76.5GHz Automotive Radar Applications with 55dB IF Channel-to-Channel Isolation," in *Proc. IEEE European Microwave Integrated Circuits Conference (EuMIC)*, Sep. 2009, pp. 192–195.

[50] Yang, S. J., et al., "A Low Noise, High Gain, Highly Linear Mixer for 77 GHz Automotive Radar Applications in SiGe:C Bipolar Technology," in *Proc. IEEE European Solid-State Circuit Conference (ESSCIRC)*, Sep. 2009, pp. 312–315.

8

Conclusions

Recent advances in technologies, circuits, and architectures of highly integrated low-power radars have been presented in the book. Single-chip or single-board implementations of the main radar subblocks have been also presented with reference to application case studies in the field of vital signs contactless monitoring, harbor traffic control, and short-range and long-range automotive driving assistance. The main idea behind the book is that a radar, properly designed to minimize its power consumption, size, and cost, has the potentiality to become in the near future a ubiquitous contactless sensor for large-market applications.

The large business that can derive from ubiquitous radar sensing applications justifies research from industry and academia in this direction. According to forecast market analysis, 7% of all vehicles sold in 2014 worldwide, particularly in Europe, Japan, and the United States, will be equipped with radar systems with a market value of several billions of U.S. dollars. In a few years, this value can grow, becoming on the order of 10% to 20% of new cars having at least two different radar devices (one short range covering the rear part of the car and one long range covering the front part of the car) mounted on each vehicle. This market forecast can become real if the cost for the customer of an automotive radar will be less than US$1,000. Thus, radars will be mounted not only on top models of premium car brands; they can become one of the key components for medium cars. As has happened in the past for other safety systems like antilock braking systems, in the future a radar for active safety can become a mandatory component for all cars sold in Europe, the United States, or Japan. Similar or even higher business volumes can arise from the use of low-power radars for biomedical applications (e.g., contactless vital signs monitoring for heart and breath rates).

Hereafter we review the key characteristics that make radars a unique sensor solution with respect to other competing technologies.

Unlike other contactless sensing technology (e.g., lasers, visible or infrared cameras, or ultrasounds), a radar is a sensor capable of operating in all weather conditions and gives to the user multiple bits of information and data about the sensed target: whether or not there is a target, its distance, its relative speed, and its direction of arrival. Moreover a radar works with nonionizing radiations and hence can be used for vital sign contactless monitoring in biomedical applications. Often, it can operate in harsh environment conditions (extreme temperatures, humidity, radiation levels, and so on).

To reach this goal the approach to radar design and the performances required of the radar should be different than traditional ones.

The development of conventional radar technique was mainly pushed by military applications during the Second World War with high-power, large size, and long-distance radars. Also when adopted for civil applications (e.g., long-range airborne or maritime radar for obstacle detection), a radar is typically a complex system where the design driver is the optimization of its performance rather than the minimization of its cost, size, weight, and power consumption.

As a consequence, at the state of the art, radar is a niche market technology for professional applications. Maximizing the achievable performance a "traditional" radar typically relies on high-performance devices, often custom designed, mixing different implementation technologies, since for each radar subsystem the most suited technology is used. This approach leads to high design and time costs.

At the state-of-the-art a radar is typically realized as the interconnection of a set of multiple circuit boards, each dedicated to a specific subsystem: RF or microwave front-ends with antennas, passives, switches, often with different boards for transmitters and receivers; analog to digital (A/D) and digital to analog (D/A) converter boards; baseband signal processing boards; power supply boards; and user interface boards.

As discussed in Chapter 2, the analog front-end electronics for traditional radar relies on high-performance III-V devices such as HEMT or HBT in GaAs or InP materials. These technologies allow for high-frequency applications, maximizing circuit metrics such as noise figure, amplifier gain and linearity, and transmitter power. Adopting high-end performance devices and circuits allows for optimal radar performance at the system level (target distance, sensitivity, false alarm rate, and so on). However, the cost and design time for III-V based technologies is much higher than for standard silicon transistor solutions such as CMOS and can be sustained only for professional niche markets (e.g., defense).

Similarly, as discussed in Chapter 3, for the baseband digital signal processing of conventional radars, high-performance computing platforms such as GPUs or ASIC are often adopted at the state of the art. Unfortunately, their costs and power consumption are much higher than embedded platforms used in consumer electronics.

In this book we have discussed how a different approach can be followed for radar design: similarly to other large-market electronic design solutions, standard and commercial devices can be used for a radar. The trend should be using the same technology for all radar subsystems, and the aim of the designer should be not maximizing the radar performance but optimizing the tradeoff between achievable performances and the relevant cost, size, and power consumption.

Obviously, an integrated low-power radar aims at system-level performance that is quite different from that of conventional radars. The maximum distance to reach can be less than 1 meter (e.g., in the case of contactless measure of heart rate or breath rate) to a maximum of 100–200m (long-range automotive radar for automatic cruise control).

The transmitted power is often below 20 dBm. Indeed, the power spectral density allowed by automotive radar standards is below -41 dBm/MHz around the 24-GHz portion of the spectrum and less than -9 dBm/MHz in the 77–81-GHz portion of the spectrum. The cross section of the targets (that can be small yachts, cars, pedestrians, bicycles, or parts of the body in case of biomedical applications) can be from some cm^2 to few m^2.

The size of the radar should be much less than 10 cm per side so that a radar can be easily mounted behind the front cover of a car in automotive applications or can be an handheld device for other applications.

Short wavelengths should be adopted to ease the miniaturization process (e.g., the wavelength is only 2.9 mm at 77 GHz). Indeed at wavelengths of few millimeters there is potential for high miniaturization, even for the antenna integration. The trend is designing radar systems in the following spectrum portions: The spectrum portion between 77 GHz and 81 GHz is suited for long-range and short-range radars. The 24-GHz spectrum is reserved mainly for short range radars.

As seen in Chapters 4 and 5, an analysis at the algorithmic level is not required for a revolution of state-of-the-art radar signal processing; rather, what is required is a selection and optimization of known pulsed or continuous-wave techniques for application scenarios where the final aim is not achieving the maximum possible performances but the reaching sufficient performances to provide a service with low cost, low power consumption, and small size. Hence, at the system and algorithmic levels, what is mainly required is a change of the approach of the radar design team: designing radar signal processing algorithms that are complexity aware (i.e., their computational cost and memory requirements should be a tradeoff between the system radar performance and the hardware and software capability of the implementing platform that has constrained budgets in terms of memory, number, and types of operation executable in real time, power consumption, and cost).

Mixing this proper algorithmic radar design with recent advances in silicon technologies and electronic design methods make possible the realization of highly-integrated radars with low cost, compact size, and low power consumption with achievable performances that can be enough for a large set of applications such as automotive short-range radars (SRR) for car parking, side-crash warning, collision warning, blind-spot detection, and stop and go control in urban scenarios; automotive LRR for adaptive cruise control (ACC); SRR for contactless heart and pulmonary monitoring in e-health applications; SRR for vital signs detection in case of natural disasters or war scenarios; distance measurements in industrial automation; millimeter-wave body scanners for security (e.g., in airports and banks); and networks of low-cost radars for traffic control in railway crossing or in a small harbor for small yachts.

As demonstrated by the circuital implementations discussed in Chapters 6 and 7, a single-board radar implementation is already possible.

Particularly Chapter 6 presented some circuital solutions for a wearable radar, called a noninvasive continuous heart monitoring (NIHM). The NIHM for vital function monitoring is based on a pulsed-radar device, working in the 3–10-GHz UWB spectrum, with a correlator-type receiver avoiding the use of a dedicated DSP hardware unit. This radar is completely integrated in a single-chip in standard CMOS silicon technology and has been tested for contactless measurements of heart rate and breath rate at distances in the order of 1m.

Chapter 6 also discussed a single-board implementation of FMCW radar for harbor infomobility (HI), realized using MIC technology for passive devices (antennas, filters, and matching networks) and commercial off-the-shelf components for active devices.

Chapter 7 presented integrated circuits solutions for a UWB 24-GHz radar for short-range automotive applications and 77-GHz FMCW radar for long-range applications. All the circuits presented in this chapter are integrated in a silicon-germanium (SiGe) BiCMOS technology.

In the near future, the continuous evolution in semiconductor technologies will allow the low-cost and low-power integration of a millimeter-wave radar transceiver and DSP unit in the same CMOS chip. Therefore, we foresee the future migration of low-power integrated radar from single-board radar to single-package or even single-chip systems. For short-range radars operating at several tens of GHz, the on-chip integration of the antenna will become feasible.

List of Acronyms

2D	two dimensional
3D	three dimentional
ABS	anti-lock braking system
A-MOS	accumulation MOS
ACC	adaptive or autonomous cruise control
AD	analog digital
ADAS	automotive driver assistance system
ADC	analog digital converter
ADS	Advanced Design System
AF	amplifier and filter
AGC	automatic gain control
AGU	automatic address generation
ALU	arithmetic logic unit
ASIC	application-pecific integrated circuit

BB	baseband
BE	broadband engine
BEOL	back-end-of-line
BPF	band pass filter
bpm	beats per minute
BPSK	binary phase shift keying
BiCMOS	bipolar complementary MOS
BJT	bipolar junction transistor
BV	breakdown voltage
CA	collision avoidance
CA-CFAR	cell averaging constant false alarm rate
CAD	computer-aided design
CAN	controller area network
CAS	cascode
CB	common base
CE	common emitter
CFAR	constant false alarm rate
CISC	complex instruction set computer
CMFB	common mode feedback circuit
CMP	chemical/mechanical planarization
CMOS	complementary MOS

CP	charge pump
CPI	coherent period of integration (or clock per instruction in Chapter 3)
CPS	coplanar stripline
CPW	coplanar waveguide
CU	control unit
CUDA	compute unified device architecture
CUT	cell under test
CW	continuous wave
DA	digital analog
DAC	digital analog converter
DBF	digital beamforming
DCM	digital clock manager
DDC	digital down conversion
DDR	double data rate
DDS	digital direct synthesizer
DFT	discrete Fourier transform
DLL	delay locked loop
DMA	direct memory access
DMIPS	Dhrystone millions of instructions per second
DOA	direction of arrival
DPU	data processing unit

DRAM	dynamic RAM
DSP	digital signal processing
DUC	digital up conversion
ECL	emitter coupled logic
EDAC	error detection and correction
EEPROM	electrical erasable programmable ROM
EH	e-health
EIRP	equivalent isotropic radiated power
EIRPD	equivalent isotropic radiated power density
EM	electromagnetic
ENOB	effective number of bits
EP	execution packet
ESA	European Space Agency
ESD	electrostatic discharge
ETSI	European Telecommunications Standards Institute
FCC	Federal Communication Commission
FDFIr	frequency domain FIR
FDSOI	full-depleted silicon on insulator
FET	field-effect transistor
FIR	finite impulse response
FFT	fast Fourier transform

FF	flip flop
FLMCW	frequency linearly modulated continuous waveform
FLOPS	floating-point operation per second
FPGA	field programmable gate array
FPSoC	field programmable system-on-chip
FSR	self-resonance frequency
GaAs	gallium arsenide
GaN	gallium nitride
GMACs	giga multiply-and accumulate operations per seconds
GO-CFAR	greatest of CFAR
GOPS	giga operations per second
GP	ground plane
GPP	general-purpose processor
GPU	graphics processing unit
GUI	graphical user interface
HBT	heterojunction bipolar transistor
HDL	Hardware Description Language
HEMT	high electron mobility transistor
HI	harbor infomobility
HPA	high-power amplifier
HPF	high-pass filter

HS	harbor surveillance
HW	hardware
IC	integrated circuit
IF	intermediate frequency
IFFT	inverse fast Fourier transform
iid	independent and identically distributed
IIR	infinite impulse response
InP	indium phosphide
I/O	input-output
IP	intellectual property
IIP1dB	input-returned 1-dB compression point
I/Q	In-phase/quadrature
ISM	industrial scientifical medical
ITAR	International Traffic Army Regulation
LASER	light amplification by stimulated emission of radiation
LE	logic element
LIDAR	light detection and ranging
LISA	Language for Instruction Set Architectures
LNA	low-noise amplifier
LP	low-pass
LPF	low-pass filter

LO	local oscillator
LRR	long-range radar
LTCC	low-temperature cofired ceramic
LUT	look-up table
MBPS	megabit per second
MAC	multiply and accumulate
MCM	multichip module
MCU	microcontroller unit
MEMS	microelectromechanical system
MIC	microwave integrated circuit
MIM	metal-insulator-metal
MIMO	multiple-input multiple-output
ML	maximum likelihood
MMIC	monolithic microwave integrated circuit
mmW	millimeter wave
MOM	metal-oxide-metal
MOS	metal-oxide semiconductor
MOSFET	metal–oxide–semiconductor field-effect transistor
MPSOC	multiprocessor system-on-chip
MRR	medium-range radar
MSPS	megasamples per second

MUSIC	multiple signal classification
NIHM	noninvasive continuous health monitoring
NF	noise figure
NCO	numerical controlled oscillator
NOP	no operation
OIP1dB	output-referred 1-dB compression point
OS-CFAR	ordered statistic, CFAR
PA	power amplifier
PAE	power-added efficiency
PC	pulse compression
PCB	printed circuit board
PCI	peripheral computer interface
Pd	detection probability
PD	power dissipation
PFD	phase frequency detector
PG	pulse generation
PGA	programmable gain amplifier
PGS	polysilicon patterned ground shield
PLL	phase-locked loop
PPE	power processor element
PR	pulse repetition

PRI	pulse repetition interval
PRF	pulse repetition frequency
PROM	programmable ROM
PSD	power spectral density
PVT	process/voltage/temperature
QFN	quad flat no-lead
RAM	random access memory
RADAR	radio detection and ranging
R&D	research and development
RISC	reduced instruction set computer
RF	radio frequency
ROM	read-only memory
RTL	register transfer level
RTOS	real-time operating system
RX	receiver
SEC/DED	single error correction/double error detection
SDE	software development environment
Si	silicon
SiC	silicon carbide
SiGe	silicon germanium
SIMD	single instruction multiple data

SIP	system-in-package
SL	sequential lobing
SMD	surface-mounting device
SNR	signal-to-noise ratio
SoC	system-on-chip
SO-CFAR	smallest of CFAR
SOI	silicon-on-insulator
SP	synergistic processor
SR	short-range
SRAM	static RAM
SRR	short-range radar
STAP	space-time adaptive processing
SW	software
TCR	transformer characteristic resistance
TDFIR	time domain FIR
TH	time hopping
Tint	integration time
TM-CFAR	trimmed mean CFAR
TMR	triple module redundancy
TOF	time of flight
ToT	time on target

TPG	triangular pulse generator
TRG	trigger
TX	transmitter
USD	United States dollars
UWB	ultrawideband
VCO	voltage-controlled oscillator
VGA	variable gain amplifier
VHDL	Very-High-Speed Integrated Circuit Hardware Description Language
VLIW	very long instruction word

About the Authors

Maria S. Greco graduated in electronic engineering in 1993 and received a Ph.D. degree in telecommunication engineering in 1998 from University of Pisa, Italy. From December 1997 to May 1998 she was with the Georgia Tech Research Institute, Atlanta, GA as a visiting research scholar, where she carried out research activity in the field of radar detection in non-Gaussian background.

In 1993 she joined the Department of Information Engineering at the University of Pisa, where she has been an associate professor since December 2011. She has been an IEEE fellow since January 2011 and she was corecipient of the 2001 IEEE Aerospace and Electronic Systems Society's Barry Carlton Award for Best Paper and recipient of the 2008 Fred Nathanson Young Engineer of the Year award for contributions to signal processing, estimation, and detection theory. She has been cogeneral-chair of the 2007 International Waveform Diversity and Design Conference (WDD07), Pisa, Italy; on the Technical Committee of the 2006 EURASIP Signal and Image Processing Conference (EUSIPCO), Florence, Italy; on the Technical Committee of the 2008 IEEE Radar Conference, Rome, Italy; on the Organizing Committee of the CAMSAP09, technical cochair of CIP2010 (Elba Island, Italy) and general co-chair of CAMSAP2011 (San Juan, Puerto Rico). She was guest coeditor of the special issue of the *Journal of the IEEE Signal Processing Society on Special Topics in Signal Processing* on "Adaptive Waveform Design for Agile Sensing and Communication" published in June 2007 and lead guest editor of the special issue of *International Journal of Navigation and Observation* on "Modeling and Processing of Radar Signals for Earth Observation" published in August 2008. She is the associate editor of both *IET Proceedings—Sonar, Radar, and Navigation* and the *IEEE Transactions on Signal Processing*, associate editor-in-chief of the *IEEE Aerospace and Electronic Systems* magazine, a member of the editorial board of the Springer *Journal of Advances in Signal Processing*, and a member of the IEEE

Signal Processing Theory and Methods and Signal Array Processing Technical Committees. She is also a member of the IEEE AES Board of Governors and vice-chair of the IEEE AES Radar Panel.

She is a coauthor of the tutorials entitled "Radar Clutter Modeling" presented at the International Radar Conference (May 2005, Arlington, VA) and "Sea and Ground Radar Clutter Modeling" presented at the 2008 IEEE Radar Conference (May 2008, Rome, Italy) and at the 2012 IEEE Radar Conference (May 2012, Atlanta, GA), as well as coauthor of the tutorial "RF and Digital Components for Highly Integrated Low-Power Radar" presented at the same conference.

Her general interests are in the areas of statistical signal processing, estimation, and detection theory. In particular, her research interests include clutter models, spectral analysis, coherent and incoherent detection in non-Gaussian clutter, CFAR techniques, radar waveform diversity, and bistatic/mustistatic active and passive radar. She coauthored many book chapters and more than 140 journal and conference papers.

Bruno Neri was born in 1956 and received his Laurea degree cum laude from the University of Pisa in 1980. In 1983 he joined the Department of Information Engineering of the same university, where he has been professor of electronics since 2000. In recent years, his research activity has addressed the design of radio-frequency integrated circuits for mobile communications and for biomedical applications. From 2002 to 2010 he was director of the Department of Information Engineering at the University of Pisa. Presently, he is a technical consultant in the lawsuit for the shipwreck of the Costa Concordia cruise ship.

Bruno Neri is coauthor of more than 100 papers published in peer-reviewed journals and proceedings of international conferences.

Giuseppe Palmisano received a Laurea degree in electronic engineering from the University of Pavia, Pavia, Italy, in 1982. From 1983 to 1991, he was a researcher in the Department of Electronics at the University of Pavia, where he was involved in CMOS and BiCMOS analog integrated circuit (IC) design. In 1992, he was visiting professor at Universidad Autonoma Metropolitana (UAM), Mexico City, Mexico, where he held a course on microelectronics for Ph.D. students. In 1993 and 2000, he joined the Faculty of Engineering at the University of Catania as associate professor and full professor, respectively, teaching analog electronics.

Since 1999, he has been leading the Radio Frequency Advanced Design Center (RF-ADC), a joint research center supported by University of Catania and STMicroelectronics (ST), Catania, Italy. The research group in RF-ADC is composed of University and ST researchers who are involved in design activity of RF and microwave ICs for data communications using modern silicon and nonsilicon emerging technologies.

In the last decade, he worked on the implementation of several innovative ICs in silicon processes such as power amplifiers for mobile phone applications, satellite receivers for digital TV, RF front-ends for WLANs, ultralow-power transceivers for wireless sensor networks, microwave ICs for automotive radar sensors, and batteryless transceivers with RF energy harvesting. He was also responsible for two FP7 European research projects dealing with analog electronic design on plastic substrate with organic transistors.

In microwave design, he implemented several innovative ICs that include complete systems such as a 24-GHz UWB short-range radar sensor and a chip set solution for the radio front-ends of a 77-GHz long-range radar sensor, and several new topologies of 77-GHz power amplifiers and VCOs were also proposed. His research achievements in all these topics have appeared in several publications of the most important international journals and conference proceedings.

Giuseppe Palmisano has also been conducting his research activity within the framework of Italian and European research projects, as well as within research agreements with electronic industries, mainly STMicroelectronics. He has been using a wide variety of silicon technologies for VLSI design approaches that range from pure Si bipolar and SiGe BiCMOS to nanoscale CMOS processes. Recently, he has been working with organic TFT technologies for flexible electronics.

He is the coauthor of more than 250 papers in international journals and conference proceedings 50 international patents, and two books.

His research interests lie in the area of analog integrated circuit design with special attention to RF and microwave applications. Recently, he has been working with ultralow-power sub-GHz RF front ends, RF remotely powered transceivers, and wireless power/data transfer with galvanic isolation.

Egidio Ragonese received a Laurea degree in electrical engineering (magna cum laude) and a Ph.D. in electronics and automation engineering from the University of Catania, Italy, in 1999 and 2003, respectively. From 2003 to 2011, he was with the Dipartimento di Ingegneria Elettrica Elettronica e Informatica (DIEEI) of the University of Catania, holding the position first of research fellow and then research assistant professor. He also was adjunct professor at the University Kore of Enna, Italy, as well as at the Scuola Superiore (Mediterranean University Center) of Catania, teaching electronics for graduate and postgraduate programs. Since 1999, while working in academia, he was deeply involved in industrial R&D programs within the Radio Frequency Advanced Design Center (RF-ADC), a joint research center supported by the University of Catania and STMicroelectronics, Catania, Italy. In October 2011, he joined STMicroelectronics, Catania, Italy, where he is RF senior designer/project leader in the R&D department.

The research of Egidio Ragonese was aimed at the conception, design, and experimental characterization of RF and mm-wave integrated circuits/systems for wireless applications (UMTS, W LAN, UWB, DVB-S, vehicular radar, mm-wave imaging, RF harvesting, RFID, wireless power transfer, etc.) in different platforms and technological process (CMOS, SiGe BiCMOS, organic TFTs, etc.). This activity was mainly developed within the framework of Italian or European research projects as well as joined academia-industrial R&D programs for which Dr. Ragonese was in charge of the technical/project management. From 2006 to 2011, he worked within the mm-wave research team of the RF-ADC to develop first 24-GHz and then 77-GHz radar sensors under joined research programs between the University of Catania and the Automotive Product Group of STMicroelectronics.

Since January 2012 Dr. Ragonese has been an IEEE senior member. He is also a member of the editorial board of Active and Passive Electronic Components, Hindawi Publishing Corporation. He served as a member of the reviewing board of several IEEE and IET journals and international conferences.

Dr. Ragonese is the coauthor of more than 60 papers in international journals and conference proceedings, several patents, and a book on integrated inductors and transformers for RF/mm-wave applications.

Sergio Saponara, IEEE senior member, got his master's and Ph.D. degrees in electronic engineering from the University of Pisa. Since 2002 he has been a Marie Curie Research Fellow at IMEC, Leuven, Belgium.

He teaches courses in vehicular electronics, electronic systems for robotics, and embedded systems at Pisa University. He also teaches fundamentals of electronics at the Italian Naval Academy in Livorno.

He is member of two IEEE standardization activities as well as the technical committees TC-7 and TC-19 of the IEEE Instrumentation and Measurement Society, and he served in the organization of several IEEE and SPIE conferences and as guest editor of special issues in several ISI journals.

Saponara has served as a member of the reviewing board of several IEEE, IET, Elsevier, and Springer journals. He held plenary or invited talks and tutorials at IEEE and SPIE conferences (e.g., the plenary talk at IEEE Radar Conference 2012). He is also associate editor of the Springer *Journal of Real-Time Image Processing*. He is member of the engineering Ph.D. school "Leonardo da Vinci" at Pisa University. His research interests include embedded computing engineering, integrated electronics, and applied electronic system..

He has coauthored more than 200 scientific publications and 10 patents. He got best paper awards at PRIME 2007, StreamingDay2010, GE2007, and second prize at the Element-14 Medical Design Award 2011. An abstract of his Ph.D. thesis was awarded at the first EDAA Ph.D. forum in Munich, 2003, and he received the "premio giovane ricercatore" from the Senato Accademico-University of Pisa.

His technology transfer activity includes collaborations with STMicroelectronics, Renesas, Selex, Ericsson, Piaggio, Magna, AustriaMicroSystems, CAEN, Sitael, Intecs, Rico, Flyby, and IDS. He served as project manager or WP manager on several national and international research projects (including PRIN, FIRB, POR-CREO, ESA, INFN, FP6, and FP7).

He is an associate member of the Italian National Institutes for Nuclear Physics and of the National Inter-University Consortium for Telecommunications. He also joined the NEWCOM and HIPEAC European networks of excellence.

Index

Recent Titles in the Artech House Radar Series

David K. Barton, Series Editor

Adaptive Antennas and Phased Arrays for Radar and Communications, Alan J. Fenn

Advanced Techniques for Digital Receivers, Phillip E. Pace

Advances in Direction-of-Arrival Estimation, Sathish Chandran, editor

Airborne Pulsed Doppler Radar, Second Edition, Guy V. Morris and Linda Harkness, editors

Bayesian Multiple Target Tracking, Second Edition Lawrence D. Stone, Roy L. Streit, Thomas L. Corwin, and Kristine L Bell

Beyond the Kalman Filter: Particle Filters for Tracking Applications, Branko Ristic, Sanjeev Arulampalam, and Neil Gordon

Cognitive Radar: The Knowledge-Aided Fully Adaptive Approach, Joseph R. Guerci

Computer Simulation of Aerial Target Radar Scattering, Recognition, Detection, and Tracking, Yakov D. Shirman, editor

Design and Analysis of Modern Tracking Systems, Samuel Blackman and Robert Popoli

Detecting and Classifying Low Probability of Intercept Radar, Second Edition, Phillip E. Pace

Digital Techniques for Wideband Receivers, Second Edition, James Tsui

Electronic Intelligence: The Analysis of Radar Signals, Second Edition, Richard G. Wiley

Electronic Warfare in the Information Age, D. Curtis Schleher

Electronic Warfare Target Location Methods, Second Edition, Richard A. Poisel

ELINT: The Interception and Analysis of Radar Signals, Richard G. Wiley

EW 101: A First Course in Electronic Warfare, David Adamy

EW 102: A Second Course in Electronic Warfare, David Adamy

EW 103: Tactical Battlefield Communications Electronic Warfare, David Adamy

Fourier Transforms in Radar and Signal Processing, Second Edition, David Brandwood

Fundamentals of Electronic Warfare, Sergei A. Vakin, Lev N. Shustov, and Robert H. Dunwell

Fundamentals of Short-Range FM Radar, Igor V. Komarov and Sergey M. Smolskiy

Handbook of Computer Simulation in Radio Engineering, Communications, and Radar, Sergey A. Leonov and Alexander I. Leonov

High-Resolution Radar, Second Edition, Donald R. Wehner

Highly Integrated Low-Power Radars, Sergio Saponara, Maria Greco, Egidio Ragonese, Giuseppe Palmisano, and Bruno Neri

Introduction to Electronic Defense Systems, Second Edition, Filippo Neri

Introduction to Electronic Warfare, D. Curtis Schleher

Introduction to Electronic Warfare Modeling and Simulation, David L. Adamy

Introduction to RF Equipment and System Design, Pekka Eskelinen

Introduction to Modern EW Systems, Andrea De Martino

The Micro-Doppler Effect in Radar, Victor C. Chen

Microwave Radar: Imaging and Advanced Concepts, Roger J. Sullivan

Millimeter-Wave Radar Targets and Clutter, Gennadiy P. Kulemin

Modern Radar Systems, Second Edition, Hamish Meikle

Modern Radar System Analysis, David K. Barton

Modern Radar System Analysis Software and User's Manual, Version 3.0, David K. Barton

Monopulse Principles and Techniques, Second Edition, Samuel M. Sherman and David K. Barton

MTI and Pulsed Doppler Radar with MATLAB®, Second Edition, D. Curtis Schleher

Multitarget-Multisensor Tracking: Applications and Advances Volume III, Yaakov Bar-Shalom and William Dale Blair, editors

Precision FMCW Short-Range Radar for Industrial Applications, Boris A. Atayants, Viacheslav M. Davydochkin, VIctor V. Ezerskiy, Valery S. Parshin, and Sergey M. Smolskiy

Principles of High-Resolution Radar, August W. Rihaczek

Principles of Radar and Sonar Signal Processing, François Le Chevalier

Radar Cross Section, Second Edition, Eugene F. Knott, et al.

Radar Equations for Modern Radar, David K. Barton

Radar Evaluation Handbook, David K. Barton, et al.

Radar Meteorology, Henri Sauvageot

Radar Reflectivity of Land and Sea, Third Edition, Maurice W. Long

Radar Resolution and Complex-Image Analysis, August W. Rihaczek and Stephen J. Hershkowitz

Radar Signal Processing and Adaptive Systems, Ramon Nitzberg

Radar System Analysis, Design, and Simulation, Eyung W. Kang

Radar System Analysis and Modeling, David K. Barton

Radar System Performance Modeling, Second Edition, G. Richard Curry

Radar Technology Encyclopedia, David K. Barton and Sergey A. Leonov, editors

Radio Wave Propagation Fundamentals, Artem Saakian

Range-Doppler Radar Imaging and Motion Compensation, Jae Sok Son, et al.

Robotic Navigation and Mapping with Radar, Martin Adams, John Mullane, Ebi Jose, and Ba-Ngu Vo

Signal Detection and Estimation, Second Edition, Mourad Barkat

Signal Processing in Noise Waveform Radar, Krzysztof Kulpa

Space-Time Adaptive Processing for Radar, J. R. Guerci

Special Design Topics in Digital Wideband Receivers, James Tsui

Theory and Practice of Radar Target Identification, August W. Rihaczek and Stephen J. Hershkowitz

Time-Frequency Signal Analysis with Applications, Ljubiša Stanković, Miloš Daković, and Thayananthan Thayaparan

Time-Frequency Transforms for Radar Imaging and Signal Analysis, Victor C. Chen and Hao Ling